THE FRONTIERS COLLECTION

THE FRONTIERS COLLECTION

Series Editors:
A.C. Elitzur M.P. Silverman J. Tuszynski R. Vaas H.D. Zeh

The books in this collection are devoted to challenging and open problems at the forefront of modern science, including related philosophical debates. In contrast to typical research monographs, however, they strive to present their topics in a manner accessible also to scientifically literate non-specialists wishing to gain insight into the deeper implications and fascinating questions involved. Taken as a whole, the series reflects the need for a fundamental and interdisciplinary approach to modern science. Furthermore, it is intended to encourage active scientists in all areas to ponder over important and perhaps controversial issues beyond their own speciality. Extending from quantum physics and relativity to entropy, consciousness and complex systems – the Frontiers Collection will inspire readers to push back the frontiers of their own knowledge.

Other Recent Titles

For a complete list of titles in The Frontiers Collection, see back of book

Joe Rosen

SYMMETRY RULES

How Science and Nature Are Founded on Symmetry

With 86 Figures and 4 Tables

 Springer

Dr. Joe Rosen
338 New Mark Esplanade
Rockville, MD 20850-2734, USA
e-mail: joerosen@mailaps.org

Series Editors:

Avshalom C. Elitzur
Bar-Ilan University,
Unit of Interdisciplinary Studies,
52900 Ramat-Gan, Israel
email: avshalom.elitzur@weizmann.ac.il

Mark P. Silverman
Department of Physics, Trinity College,
Hartford, CT 06106, USA
email: mark.silverman@trincoll.edu

Jack Tuszynski
University of Alberta,
Department of Physics, Edmonton, AB,
T6G 2J1, Canada
email: jtus@phys.ualberta.ca

Rüdiger Vaas
University of Gießen,
Center for Philosophy and Foundations of Science
35394 Gießen, Germany
email: Ruediger.Vaas@t-online.de

H. Dieter Zeh
University of Heidelberg,
Institute of Theoretical Physics,
Philosophenweg 19,
69120 Heidelberg, Germany
email: zeh@uni-heidelberg.de

Cover figure: Image courtesy of the Scientific Computing and Imaging Institute,
University of Utah (www.sci.utah.edu).

ISBN 978-3-642-09508-5   e-ISBN 978-3-540-75973-7

DOI 10.1007/978-3-540-75973-7

Frontiers Collection ISSN 1612-3018

Cover design: KünkelLopka, Werbeagentur GmbH, Heidelberg, Germany

Printed on acid-free paper

9 8 7 6 5 4 3 2 1

springer.com

For Mira

Preface

Ernest Rutherford (New Zealand–British physicist, 1871–1937), the 1908 Nobel Laureate who discovered the existence of atomic nuclei, is famously quoted as having said: "Physics is the only real science. All the rest is butterfly collecting." Or something to that effect. I like to include this quote in my introductory remarks at the first class meetings of the physics courses I teach.

I have seen that there are those who interpret this as a put-down of amateurs (butterfly collectors) in science. However, my own interpretation of Rutherford's statement is that he is claiming that, except for physics, all of the rest of science is involved merely in collecting facts and classifying them (butterfly collecting). It is physics, unique among the sciences, that is attempting to find *explanations* for the classified data.

The periodic table of the chemical elements, originally proposed by Dmitri Ivanovich Mendeleev (Russian chemist, 1834–1907), presents an example of this. Chemists toiled to discover the chemical elements and their properties and then classified the elements in the scheme that is expressed by the periodic table. Here was the chemists' butterfly collecting. It took physicists to *explain* the periodic table by means of quantum theory.

Rutherford's assessment of science might well have held a large degree of validity in the 19th and early 20th centuries. But since then other fields of science than physics have developed 'physics envy' and they too are now busy searching for explanations. For example, chemistry finds its explanations in physics. And explanations in biology are found, on one level, in evolution theory and, on another level, in chemistry and physics.

I differ with Rutherford, though, in his narrow conception of science. To be sure, science involves searching for explanations. But production and collection of data through experimentation and observation and classification of the data supply the raw material for science to attempt to explain. Without them there would be nothing to explain and no 'real science' in Rutherford's sense. So I include butterfly collecting in my broad conception of science.

The point of all that, for the purpose of this book, is to lead to the notion that science – even in its broad conception – not only makes much use of symmetry, but is essentially and fundamentally based on symmetry. Indeed, science rests firmly on the triple foundation of reproducibility, predictability, and reduction, all of which are symmetries, with additional support from analogy and objectivity, which are symmetries too. So it is not much of an exaggeration to claim that science is symmetry. Or perhaps in somewhat more detail, science is our view of nature through symmetry spectacles. That is one component of the main thesis of this book.

In addition to an exposition and justification of this central idea, that science is founded on symmetry, we also look into how symmetry is used in science in general and in physics in particular (Rutherford's 'real science'). And we find: symmetry of evolution (symmetry of the laws of nature), symmetry of states of physical systems, gauge symmetry of the fundamental interactions, and the symmetry inherent to quantum theory. So not only do we *view* nature through symmetry spectacles, but we *understand* nature in the language of symmetry. That is another component of this book's main thesis.

All that leads to deep questions that await clarification. What is the source of all this symmetry? What is nature telling us? Is nature symmetry, at least in some sense? If not at the level that physics is presently investigating, are deeper levels of reality involved with symmetry in a very major way? Or even, will symmetry turn out to be what those fundamental levels are *all* about? Is symmetry the foundational principle of the Universe?

Such ideas lurk in the back of many physicists' minds, and some physicists express them outrightly. Brian Greene, for one, states in Chap. 8 of [1]: "From our modern perspective, symmetries are the foundation from which laws spring." And Stenger [2] adds his vote.

Speaking of the Universe, it is shown in this book that the Universe cannot possess exact symmetry. This connects to conceptual problems with symmetry breaking at 'phase transitions' in the evolution of the

Universe according to big-bang type cosmological schemes. Such and related matters are discussed, including the nature of the 'quantum era' that is assumed to form the first evolutionary stage in big-bang type schemes. But many questions remain for future elucidation. Are big-bang type cosmological schemes the best models for the evolution of the Universe? If so, did the Universe pass through distinct eras separated by transitions that might be characterized as 'phase transitions'? What were the properties of the eras and of the transitions? Was there a 'quantum era'? If there was, can it be meaningfully described? And can present-day high-energy physics reflect the properties of earlier stages in the evolution of the Universe? If it can, what will the results of experiments soon to be performed at high-energy laboratories, such as CERN's Large Hadron Collider, reveal about the earlier Universe? And what will they tell us about today's physics? Will they help clarify or will they sow confusion?

Here is the order of presentation: We start in Chap. 1 with a brief introduction to the concept of symmetry, including an analysis of the intimate relation between symmetry and asymmetry – especially that symmetry implies asymmetry – and a discussion of analogy and classification as symmetry. We then see in Chap. 2 what science is, how it makes use of symmetry, and how it is based solidly on symmetry. So solidly, in fact, that one might well view science as symmetry. In Chap. 3 we consider a number of ways in which physics, in particular, additionally makes use of symmetry. Since physics underlies the other sciences, we find that science is based even more solidly on symmetry, and perhaps nature will turn out to possess a symmetry foundation as well. The symmetry principle, also known as Curie's principle, is derived in its various versions in Chap. 4. We see in Chap. 5 two ways in which the symmetry principle is very usefully applied in science. In Chaps. 6 and 7 we discuss the ideas of imperfect symmetry and symmetry in general and as applied to the Universe and its evolution, as well as related ideas.

There then follows the more formal part of the book, in which we develop a formalism of symmetry. Chapters 8 and 9 form a brief introduction to group theory, the mathematical language of symmetry, which is indispensable for serious quantitative, as well as qualitative, applications of symmetry in science, mostly in physics and chemistry. Nevertheless, in spite of that indispensability, Chaps. 8 and 9 can be skipped without too much harm to those preferring a more conceptual approach. Chapter 10 develops the language and formalism that underlie the application of symmetry. Group theory is unavoidable there, but I try to allow the reader to make sense of the ideas even without group theory. And finally, in Chap. 11 we apply symmetry considerations and the symmetry formalism to physical processes and derive the symmetry principles that apply to them.

Chapter 12 brings together and summarizes the principles of symmetry that are developed and presented in this book.

I would like to express my thanks to my friends and colleagues Avshalom C. Elitzur and Lawrence W. Fagg, who kindly read the manuscript of this book and helped me with their comments and suggestions. And especially, I thank my wife, Mira Frost, for her unflagging support and for putting up with my disappearances into my study to work on this book.

Rockville, Maryland,   *J. Rosen*
August 2007

Contents

1

The Concept of Symmetry

1.1 The Essence of Symmetry

Everyone has some idea of what *symmetry* is. We recognize the bilateral (left-right) symmetry of the human body, of the bodies of many other animals, and of numerous objects in our environment. We enjoy the rotation symmetry of many kinds of flower. We consider a scalene triangle, one with all sides unequal, to be completely lacking in symmetry, while we see symmetry in an isosceles triangle and even more symmetry in an equilateral triangle. That is only for starters. Any reader of this book can easily point out many more kinds and examples of symmetry.

In science, of course, our recognition and utilization of symmetry is often more sophisticated, sometimes very much more. But what symmetry actually boils down to in the final analysis is that *the situation possesses the possibility of a change that leaves some aspect of the situation unchanged.*

A bilaterally symmetric body can be reflected through its midplane, through the (imaginary) plane separating the body's two similar halves. Think of a two-sided mirror positioned in that plane. Such a reflection is a change. Yet the reflected body looks the same as the original one; it coincides with the original: the reflected right and left hands, paws, or hooves coincide, respectively, with the original left and right ones, and similarly with the feet, ears, and other paired parts (see Fig. 1.1).

For the triangles let us for simplicity confine ourselves to rotations and reflections within the plane of the figures. Then a rotation is made about a point in the plane, which is the point of intersection of the

Fig. 1.1. Bilateral symmetry

axis of rotation that is perpendicular to the plane. A reflection is made through a line in the plane, where the line is where a two-sided mirror that is perpendicular to the plane intersects the plane. An infinite number of such changes can be performed on any triangle. But for an equilateral triangle there are only a finite number of them that can be made on it and that nevertheless leave its appearance unchanged, i.e., rotations and reflections for which the changed triangle coincides with the original. They are rotations about the triangle's center by 120° and by 240°, and reflections through each of the triangle's three heights, five changes altogether (see Fig. 1.2). (For the present we do not count rotations by multiples of 360°, which are considered to be no change at all.)

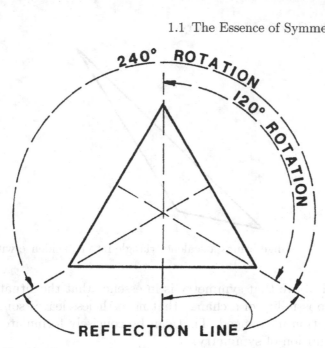

Fig. 1.2. Changes bringing an equilateral triangle into coincidence with itself

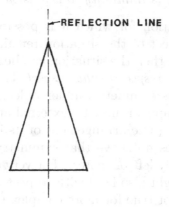

Fig. 1.3. Change bringing an isosceles triangle into coincidence with itself

Although an infinity of planar rotations and reflections can also be performed on any isosceles triangle, there is only a single such change that preserves the appearance of such a triangle, that leaves the triangle coinciding with itself. It is reflection through the height on its base (see Fig. 1.3). And a scalene triangle cannot be made to coincide with itself by any planar rotation or reflection, once again not counting rotations by multiples of 360° (see Fig. 1.4).

Fig. 1.4. No change brings a scalene triangle into coincidence with itself

I stated above that symmetry is in essence that the situation possesses the possibility of a change that nevertheless leaves some aspect of the situation unchanged. That can be concisely formulated as this precise definition of symmetry:

Symmetry is immunity to a possible change.

When we have a situation for which it is possible to make a change under which some aspect of the situation remains unchanged, i.e., is immune to the change, then the situation can be said to be *symmetric under the change with respect to that aspect.* For example, a bilaterally symmetric body is symmetric under reflection through its midplane with respect to appearance. Its external appearance is immune to midplane reflection. (The arrangement of its internal organs, however, most usually does not have that symmetry. The human heart, for instance, is normally left of center.) For very simple animals, their bilateral symmetry might also hold with respect to physiological function as well. That is not true for more complex animals.

An equilateral triangle is symmetric with respect to appearance under the rotations and reflections we mentioned above. An isosceles triangle is symmetric with respect to appearance under reflection through the height on its base. But a scalene triangle is not symmetric with respect to appearance under any planar rotation or reflection.

Note the two essential components of symmetry:

1. *Possibility of a change.* It must be possible to perform a change, although the change does not actually have to be performed.
2. *Immunity.* Some aspect of the situation would remain unchanged, if the change were performed.

If a change is possible but some aspect of the situation is not immune to it, we have *asymmetry*. Then the situation can be said to be *asymmetric under the change with respect to that aspect*. For example, a scalene triangle is asymmetric with respect to appearance under all planar rotations and reflections. All triangles are asymmetric with respect to appearance under 45° rotations. While equilateral triangles are symmetric with respect to appearance under 120° rotations about their center, isosceles triangles do not possess this symmetry; they are asymmetric under 120° rotations with respect to appearance. And while a triangle might be symmetric or asymmetric with respect to appearance under a given rotation or reflection, all triangles are symmetric under all rotations and reflections with respect to their area; rotations and reflections do not change area. On the other hand, all plane figures are asymmetric with respect to area under dilation, which is enlargement (or reduction) of all linear dimensions by the same factor. The area then increases (or diminishes) by the square of that factor.

If there is no possibility of a change, then the very concepts of symmetry and asymmetry will be inapplicable. For example, if the property of color is not an ingredient of the specification of a plane figure, then the change of, say, color interchange will not be a possible change for such a figure. Thus color interchange symmetry or asymmetry will not be conceptually applicable to the situation. Or alternatively, one might say that such a plane figure will possess trivial symmetry under such a change. One might say that all its aspects will be trivially immune to such a change. It is a matter of taste, but I tend to prefer calling it inapplicability rather than triviality.

If, however, color *is* included in the specification of a figure, then color interchange will become a possible change for it. For example, if the figure is black and white, it will be symmetric under red–green interchange with respect to appearance. Interchange red and green, and nothing will happen to the figure. If the figure is black and green, it will be asymmetric under the same change with respect to the same aspect. Interchange red and green, and the figure will become black and red, which is not the same as black and green.

As an example, consider the black–white figure of Fig. 1.5. What symmetries can we find lurking here? For simplicity let us confine ourselves to the plane of the figure, as we did earlier. If we consider only the geometric properties of the figure and ignore its coloring, then the figure possesses the symmetry of the square with respect to its appearance: it is symmetric under rotations by 90°, 180°, and 270°,

Fig. 1.5. Black–white figure as an example

and under reflections through each of its diagonals and through each of its vertical and horizontal midlines. (You might want to indicate these four lines in the figure.) That adds up to seven changes under which the figure is symmetric with respect to appearance, ignoring its coloration.

Let color enter our considerations. Some of the changes that did not change appearance before, now do make a difference. They are rotations by 90° and 270° and reflections through the vertical and horizontal midlines. That leaves the colored figure symmetric with respect to appearance only under rotation by 180° and reflections through each of the two diagonals, three changes. With color now in the picture, we can consider black–white interchange. However, the figure is asymmetric under this change. Nevertheless, we can still find symmetry under black–white interchange if we combine the interchange with a geometric change to form a compound change. Thus the figure is symmetric with respect to appearance under the compound changes consisting of black–white interchange together with rotation by 90°, with rotation by 270°, with reflection through the vertical midline, and with reflection through the horizontal midline, making four compound changes.

Approximate symmetry is approximate immunity to a possible change. There is no approximation in the change or in its possibility; it must indeed be possible to perform a change. The approximation is in the immunity. Some aspect of the situation must change by only a little, however that is evaluated, when some change is performed. Then the situation can be said to be *approximately symmetric under the change with respect to that aspect*.

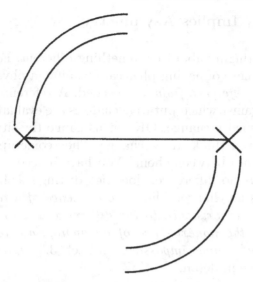

Fig. 1.6. Approximate two-fold rotation symmetry

For example, the figure of Fig. 1.6 possesses approximate two-fold rotation symmetry with respect to its appearance. Under 180° rotation about its center its appearance changes, but only by a little. And the bilateral symmetry of humans and other animals is in reality also only approximate. Not only do the internal organs not all possess that symmetry, but even for external appearance the symmetry is never exact. For instance, the fingerprints of one hand are not the mirror images of the corresponding fingerprints of the other hand, and the hand and foot of one side (usually the right side for right-handed people) are almost always slightly longer than those of the other side.

Approximate symmetry is a softening of the hard dichotomy between symmetry and asymmetry. The extent of deviation from exact symmetry that can still be considered approximate symmetry will depend on the context and the application and could very well be a matter of personal taste. The same figure, for example that of Fig. 1.6, might be considered approximately symmetric (or slightly asymmetric) by some observers, while others might consider it very asymmetric (or nowhere near symmetric). We will discuss approximate symmetry in more detail in Sect. 6.1 and will formalize the notion in Sect. 10.6.

1.2 Symmetry Implies Asymmetry

Change is the bringing about of something different. For a difference to exist, in the sense of having physical meaning, a physical gauge for the difference, a *reference frame*, is needed. A reference frame serves as a standard against which putative changes are evaluated: You think you have performed a change. OK, let us gauge the situations before and after what you think was a change. They come up different, the gauge distinguishes between them? You have indeed made a change. The gauge shows no difference, does not distinguish between them? Your 'change' is no change. Thus, *the existence of a reference frame is necessary to give existence to the difference and to the possibility of change.* And *the nonexistence of an appropriate reference frame makes a supposed change impossible.* In Sect. 3.3 we discuss the idea of reference frame in detail.

To illustrate this, think of an object floating in otherwise empty space. Now consider the change of moving the object by some distance in some direction, i.e., spatial displacement. All right, the object is now displaced, at least we might think it is. Picture the result of the move: again an object floating in otherwise empty space. Is the result of the displacement physically (as opposed to, say, philosophically) different from the original situation? No. There is no way to distinguish between the original and the displaced situations. They are identical. A displacement involves a change of location. But in otherwise empty space all locations are identical, since there is no reference frame for position in the space. So no change has taken place and the supposed change is not a possible change. The case is similar for rotation and for reflection.

In order to be capable of gauging a difference, a reference frame cannot be immune to the change that brings about the difference. It cannot be immune to the change for which it is intended to serve as reference. Otherwise it could not serve its purpose.

For example, the change of spatial displacement brings about a difference in location. A set of tape measures as coordinate axes could serve as a reference frame for that, but not if the tapes are themselves immune to displacement, i.e., not if the tapes are infinite and homogeneous (unmarked). Imagine sliding such a tape parallel to itself and compare the final and original situations of the tape with each other. There is no difference. Marked axes, which can indeed gauge differences in location and can thus serve as a reference frame for displacements, are themselves affected by displacements. Imagine sliding a marked tape, like an ordinary tape measure, parallel to itself and compare its final and initial situations with each other. They are clearly different. If you imagine sliding the tape eight meters in the positive direction, then in its final state the tape's zero mark will align with the eight meter mark of the tape in its original state.

A reference frame is a changeable aspect of a situation. Now, any changeable aspect of a situation can serve as a reference frame for that change in the situation, since it is tautological that a changeable aspect of a situation is not immune to its own change. A changeable aspect of a situation allows the possibility of a change. Indeed, we can say that it *represents* the possibility of a change and that any possibility of a change is represented by a changeable aspect of the situation. So a situation will possess symmetry if and only if it has *both* an aspect that can change – giving the possibility of a change – *and* an aspect that does not change concomitantly – giving the immunity to the possible change. In other words, the possibility of a change, which is a necessary component of symmetry, is contingent upon the existence of an asymmetry of the situation under the change. And hence the succinct result:

Symmetry implies asymmetry.

This is discussed further in [3]. Another, less succinct way of expressing the relation between symmetry and asymmetry is this:

> *Symmetry requires a reference frame, which is necessarily asymmetric. The absence of a reference frame implies identity, hence no possibility of change, and hence the inapplicability of the concept of symmetry.*

Consider, for example, the equilateral triangle of Fig. 1.2. Its appearance is an aspect of it that is immune to 120° rotation about its center in its plane, so it possesses symmetry under 120° rotation with respect to appearance. Or so we might blithely think. But what is this change we call '120° rotation'? Do what we will to the triangle – twist it, twirl it, twitch it, swivel it – when is it rotated by 120°, or rotated at all for that matter? Unless we have a reference frame to endow an orientation difference with existence and thus give significance to a rotation, all our actions amount to nil, nothing is accomplished. The situations before and after our efforts remain identical. We then have no rotation at all, nor do we have even the conceptual possibility of rotation. So the concept of rotation symmetry is inapplicable to the triangle in the absence of an appropriate reference frame.

However, the triangle is not a universe in itself. The *total* situation, that of the equilateral triangle together with its environment, does possess aspects that are not immune to 120° rotation and that can thereby serve as a reference frame for 120° rotation. The walls of the room, for instance, could serve as reference frame, since they are asymmetric under 120° rotation. Thus, rotation by 120° is indeed a change. The equilateral triangle is symmetric in the context of its environment. It is symmetric under 120° rotation thanks to its environment's lack of immunity to 120° rotation, thanks to the asymmetry of the total situation – triangle plus environment – under the rotation.

The above results concerning symmetry, change, immunity, reference frame, and asymmetry can be summarized by the following diagram, where arrows denote implication:

$$
\text{symmetry} \begin{array}{c} \nearrow \\ \searrow \end{array} \begin{array}{c} \text{possibility} \\ \text{of a change} \end{array} \longrightarrow \begin{array}{c} \text{reference for} \\ \text{the change} \end{array} \longrightarrow \begin{array}{c} \text{asymmetry} \\ \text{under the change} \end{array} \\ \text{immunity to} \\ \text{the change}
$$

Thus, *for there to be symmetry, there must concomitantly exist asymmetry under the same change that is involved in the symmetry.* For every symmetry there is an asymmetry tucked away somewhere in the Universe.

So symmetry implies asymmetry. This relation is not symmetric, since asymmetry does not imply symmetry, at least not in the same sense that symmetry implies asymmetry, in the sense that actual symmetry implies actual asymmetry, as was demonstrated above. However, asymmetry does imply symmetry in the limited sense that the lack of immunity to a possible change implies the conceptual possibility of immunity to the change. Thus actual asymmetry implies merely the conceptual possibility of, but not actual, symmetry.

1.3 Analogy and Classification Are Symmetry

A very important kind of symmetry for science, one that is not often thought of as symmetry, is *analogy*. Analogy is the immunity of the validity of a relation to changes of the elements involved in it.

To see what we actually have here, consider, for example, the relation expressed by the statement, 'An animal has a relatively long tail'. This relation involves two elements, an animal and a relatively long tail. The former element is not unique, since there are more than just a single animal in the world. Indeed, one can say, for example, 'This deer has a relatively long tail', or 'That squirrel has a relatively long tail'. But the relation is not valid for all animals. Deer do not have relatively long tails, while squirrels do. Nevertheless, there are, in fact, more than just a single animal for which the statement is valid. Squirrel A has a relatively long tail, squirrel B also has a relatively long tail, squirrel C does too, so does squirrel D, and so on for all squirrels as well as for certain other animals. The relation defines an analogy among animals: All relatively long-tailed animals are analogous in that, whatever their differences, they all possess the common property of having a relatively long tail. And as a fringe benefit we have that all relatively non-long-tailed (i.e., medium-, short-, and no-tailed) animals are analogous in that, whatever their differences, they all possess the common property that their tails are not relatively long.

We see that this analogy is symmetry by noting:

1. There is the possibility of a change. Since the relation is formulated for more than a single animal, the animal to which it is applied can be switched.
2. The validity of the relation is immune to certain such changes. The relation holds just as well for squirrel A, for squirrel B, for C, etc., who each proudly waves a relatively long tail.

For another, similar analogy consider the relation expressed by 'An astronomical body moves along an elliptical orbit with the Sun at one of its foci'. This statement, too, can be viewed as involving two elements, an astronomical body and an elliptical orbit with the Sun at one of its foci. The former element is not unique; there are more than a single astronomical body in the cosmos. For example, one can say, 'The Moon moves ...', or 'Venus moves ...'. But the relation is not valid for all astronomical bodies. The Moon does not move in such a way, while Venus does. There are, however, more than a single astronomical body for which the statement is valid. They include all the planets of the solar system, for whom the statement becomes Kepler's first law of planetary motion (Johannes Kepler, German astronomer and mathematician, 1571–1630), as well as the asteroids, dwarf planets, some of the comets, etc. The statement is an expression of analogy among astronomical bodies: All those bodies that move along elliptical orbits with the Sun at one of their foci are analogous in that, whatever their differences, they all move in just that way. And we also have that all the other astronomical bodies, such as stars and moons, are analogous in that, whatever their differences, they all do not move along ellipses with the Sun at one of the foci.

This analogy, too, is symmetry:

1. There is the possibility of a change. Since the relation is applicable to more than a single astronomical body, the body to which it is applied can be switched.
2. The validity of the relation is immune to certain such changes. The statement is valid just as well for Mars, for Neptune, for Uranus, etc., each of which moves along an elliptical orbit with the Sun at one of its foci.

Now, consider a relation involving a pair of changeable elements, 'X is the locus of all points equidistant from a given point (its center), all points lying in Y'. The relation involves a pair of elements, (X, Y), where X and Y can be any geometric objects and are certainly not unique. For instance, one can say, 'A triangle is the locus ... lying in an ellipsoid', or 'A circle is the locus ... lying in a plane'. But the relation is not valid for all pairs of geometric objects. It is not true for the pair (a triangle, an ellipsoid), while it does hold for the pair (a circle, a plane). There are more than one pair (X, Y) for which it is valid. Three of them are: (a pair of points, a line), (a circle, a plane), and (a spherical surface, space). This relation between X and Y defines an analogy among pairs of geometric objects: All those pairs (X, Y) whose elements X and Y fulfill the relation as stated are analogous in that, whatever their differences, they all fulfill the relation. And in addition, all those pairs whose elements do not fulfill the relation are analogous in that, whatever their differences, they do not fulfill the relation.

Analogies involving pairs of changeable elements are often put in the form: A is to B as C is to D as .... For the present example this form is: A pair of points is to a line as a circle is to a plane as a spherical surface is to space .... (This form of expressing an analogy can easily be generalized for relations involving any number of changeable elements.) The symmetry here is:

1. Since the relation is applicable to more than one pair of geometric objects, the pair to which it is applied can be switched.
2. The validity of the relation is immune to certain such changes.

Now, consider the experimental setup of a given sphere rolling down a fixed inclined plane, with the experimental procedure of releasing the sphere from rest, letting it roll for any time interval t, and noting the distance d the sphere rolls in this time interval. Performing n such experiments, we collect n data pairs (or data points) (t_1, d_1), ..., (t_n, d_n). The pairs obey the relation $d_k = bt_k^2$, where b is a positive proportionality constant, for $k = 1, \ldots, n$. This is a relation involving two changeable elements, as in the preceding example. The n data pairs, as well as an infinity of potential data pairs, are analogous in that they all obey the same relation, $d = bt^2$, and in that sense t_1 is to d_1 as t_2 is to d_2 as .... All other (t, d) pairs, which do not obey the relation $d = bt^2$, are also analogous in that they do not obey the relation. The symmetry is that we can switch among actual and potential data pairs, and however we switch among them, the relation between t and d remains the same.

Additional physics analogies can by found in [4].

With the help of the four examples above we now see how analogy, as the immunity of the validity of a relation under changes of the elements involved in it, is indeed what we thought we understood by the term 'analogy' before we found ourselves hopelessly confused by such a weird definition. The reason for such a definition of analogy, besides its being a good one, is that it directly exposes the symmetry that is analogy, since it implies:

1. the possibility of a change, the change of elements involved in the relation,
2. the immunity of the validity of the relation to certain such changes.

Note that analogy implies and is implied by *classification*. An analogy imposes a classification by decomposing the set of elements or element pairs, triples, etc., to which the relation is applicable into classes of analogous elements or pairs, triples, etc. For example, among animals the relation, 'An animal has a relatively long tail', separates all animals into a class of relatively long-tailed animals, those animals for which the relation is valid, and a class of relatively non-long-tailed animals, those for which the statement is false. In the astronomical example the relation, 'An astronomical body moves along an elliptical orbit with the Sun at one of its foci', decomposes the set of all astronomical bodies into a class of those for which the relation is valid, the most notable of which are the planets of the solar system, and a class of astronomical bodies that do not move according to the statement, which includes the planetary moons and all the stars, among others.

In the geometric example the relation, 'X is the locus of all points equidistant from a given point (its center), all points lying in Y', separates all pairs of geometric objects into a class of those pairs for which the relation holds, the best known of which are (a pair of points, a line), (a circle, a plane), and (a spherical surface, space), and a class of those that do not fulfill the relation, such as (a triangle, an ellipsoid) and (a hyperboloid, space). And in the laboratory example the relation $d = bt^2$ decomposes all (t, d) pairs into a class of those obeying the relation, i.e., all actual and potential data pairs for the experiment, and a class of those for which $d \neq bt^2$, those that cannot be data for the experiment.

Conversely, a classification defines the analogy of belonging to the same class. If any set is decomposed into mutually exclusive classes, then the very property of belonging to the same class defines an analogy among the elements of the set. For instance, the kids in a school can be, and for administrative purposes are, classified by grade. That makes all pupils in the same grade analogous. Or, motor vehicles can be classified by the number of axles. This classification makes all vehicles with the same number of axles analogous, which might find expression in the toll rate on toll roads.

For a detailed example, consider the classification of the chemical elements that is expressed by the periodic table, originally proposed by Mendeleev. Each column of the table comprises a group of elements possessing similar chemical properties. There is the noble gas group (helium, neon, argon, ...), the halogen group (fluorine, chlorine, bromine, ...), and so on. The analogy that is defined by this classification is the relation, 'Element X belongs to group N', where X is the changeable element for any fixed group N. Thus helium, neon, argon, ..., are analogous in that they are all noble gases. Fluorine, chlorine, bromine, ..., are analogous by all belonging to the halogen group. And so on for the other groups. The symmetry here is this [5]:

1. Every group N contains more than a single element, so the element X to which the relation 'Element X belongs to group N' is applied can be changed.
2. The validity of the relation is immune to switching among elements X that belong to group N.

This example serves also as the example of Ernest Rutherford's 'butterfly collecting' that was discussed in the introduction.

And one more example. For the purposes of blood donation and transfusion, people are classified by their blood type (A, B, AB, O) and their Rh group (+, −), giving eight classes (A+, A−, B+, ..., O−). The analogy here is among people belonging to the same blood class, with the relation, 'Individual X possesses blood of type/group Y', for changeable X and fixed Y (A+, ..., or O−). The symmetry here is:

1. More than a single person have blood of any of the eight combinations of blood type and group, so X in the relation can be changed.
2. The validity of the relation is immune to switching among people of the same blood type/group.

Thus, analogy and classification, which imply each other, are both symmetry.

1.4 Summary

Symmetry is immunity to a possible change, i.e., we have symmetry when it is possible to perform some change in the situation that nevertheless leaves some aspect of the situation unaffected. Then we have symmetry under the change with respect to that aspect. If some aspect of the situation is not immune to the change, then the situation is asymmetric under the change with respect to this aspect. Since a change requires the existence of a reference frame that is affected by the change, such a reference frame is necessarily asymmetric under the change. Thus, *for every symmetry there exists an asymmetry*. That was the gist of Sects. 1.1 and 1.2.

In Sect. 1.3 we saw that analogy is symmetry and discussed how analogy implies and is implied by classification, which is also symmetry.

Science Is Founded on Symmetry

In this chapter we briefly review what science is about, and we see that it strongly involves reduction, which is shown to be symmetry. We consider three ways reduction is used in science – observer and observed, quasi-isolated system and environment, and initial state and evolution – and see in detail the symmetry implied by each.

Reproducibility and predictability, which are both essential components of science, are shown to be symmetries as well. Since science rests firmly on the triple foundation of reproducibility, predictability, and reduction, science is solidly based on symmetry. Indeed, science can be said to *be* symmetry, at least to the extent that it is our view of nature through symmetry spectacles.

In addition, analogy, shown earlier to be symmetry, is seen to be essential for the operation of science.

2.1 Science

For the purpose of our discussion we take this definition of *nature*:

> *Nature is the material universe with which we can, or can conceivably, interact.*

The *material universe* is everything of a purely material character. Here I mean 'material' in the broad sense of anything related to matter, including such as energy, momentum, electric charge, fields, waves, and so on. To *interact* with something is to act upon it and be acted upon by it. That implies the possibility of performing observations and measurements on it and of receiving data from it, which is what we are actually interested in. To be able *conceivably* to interact with something means that, although we might not be able to interact with it at present, interaction is not precluded by any principle known to us and is considered attainable through further technological research and development. Thus nature, as the material universe with which we can, or can conceivably, interact, is everything of purely material character (in the broad sense) that we can, or can conceivably, observe and measure.

We live in nature, observe it, and are intrigued. We try to understand nature in order both to improve our lives by better satisfying our material needs and desires and to satisfy our curiosity. And what we observe in nature is a complex of phenomena, including ourselves, where *we are related to all of nature*, as is implied by our definition of nature as the material universe with which we can, or can conceivably, interact. The possibility of interaction is what relates us to all of nature and, due to the mutuality of interaction and of the relation it brings about, relates all of nature to us. It then follows that all aspects and phenomena of nature are actually interrelated, whether they appear to be so or not. Whether they are interrelated independently of us or not, they are certainly interrelated through our mediation. Thus all of nature, including *Homo sapiens*, is interrelated and integrated.

Now we come to *science*:

Science is our attempt to understand rationally and objectively the reproducible and predictable aspects of nature.

This, as we will see, is essentially the same as

Science is our attempt to understand rationally and objectively the lawful aspects of nature.

And I repeat, nature is the material universe with which we can, or can conceivably, interact. By 'our' in the above definition I mean that science is a human endeavor and is shaped by our modes of perception and our mental makeup. It is the endeavor of all humanity, not of any particular individual, so it must be as objective as possible. 'Attempt' means that we try but might not always succeed. By 'understand rationally and objectively' I mean be able to explain in a logical way that is valid for everybody. That excludes explanations based on intuition, feeling, or religious considerations, among others. We explain logically and objectively by finding *order* among the reproducible and predictable aspects of nature, formulating *laws*, and devising *theo-*

ries [6]. In the following I use 'understand' and 'explain' solely in their rational and objective sense.

Note that I am using the term 'science' in this book strictly in the sense of natural science. More specifically, I am not including such as mathematics and philosophy in what I mean by science.

Reproducibility means that experiments can be replicated in other labs and in the same lab, thus making science a common endeavor that is as objective as possible. (Reproducibility is treated in more detail in Sect. 2.3.) *Predictability* means that among the phenomena investigated, order can be found, from which laws can be formulated, predicting the results of new experiments. (See Sect. 2.4 for a more detailed treatment of predictability.) There is no claim here that all of nature's aspects are reproducible and predictable. Indeed, they are not. For example, according to our understanding of nature's quantum aspect, individual submicroscopic events, such as the radioactive decay of an unstable nucleus, are inherently unpredictable. (However, the statistical properties of many submicroscopic events, such as the half-life of a radioactive isotope, may be predictable.) And perhaps the behavior of an individual organism is inherently not completely predictable either. But all such aspects of nature lie outside the domain of concern of science. Reproducibility and predictability form two essential components of the foundation upon which science firmly rests. If either is lacking, science will be unable to operate.

As science attempts to comprehend larger- and larger-size phenomena of nature, actual reproducibility is replaced by declared reproducibility, in the sense that, even if we cannot actually replicate the effect at our pleasure, such as a volcano eruption or the birth of a star, nature supplies us with sufficient quantity and variety to enable us to investigate the phenomena. But as the size increases to truly gigantic, such as superclusters of galaxies, that reasoning becomes tenuous. Moreover, when the Universe is considered as a whole, we cannot even declare reproducibility. Whatever metaphysical ideas one might have about universes, science deals solely with the single universe we are part of, with the Universe, which is thus irreproducible by any meaning of the word. What this lengthy introduction is leading to is that the fields of cosmology (the study of the working of the cosmos, the Universe as a whole) and cosmogony (considerations of the origin of the cosmos), as fascinating as they are, cannot rightfully be considered scientific endeavors in the same sense that the more mundane branches of science are. For a more detailed discussion of this issue see Chap. 7.

2.2 Reduction Is Symmetry

But how are we to grasp the wholeness, the integrality, that is nature? When we approach nature in its completeness, it appears so awesomely complicated, due to the interrelation of all its aspects and phenomena, that it might seem utterly beyond hope to understand anything about it at all. True, some obvious simplicity stands out, such as day–night periodicity, the annual cycle of the seasons, and the fact that fire consumes. And subtler simplicity can be discerned, such as the term of pregnancy, the relation between clouds and rain, and that between the tide and the phase of the moon. Yet, on the whole, complexity seems to be the norm, and even simplicity, when considered in more detail, reveals wealths of complexity. But again due to nature's unity, any attempt to analyze nature into simpler component parts cannot but leave something out of the picture.

Holism is the approach to nature that holds that nature can be understood only in its wholeness or not at all. And this includes human beings as part of nature. As long as nature is not yet understood, there is no reason a priori to consider any aspect or phenomenon of it as being intrinsically more or less important than any other. Thus, it is not meaningful to pick out some part of nature as being more 'worthy' of investigation than other parts. Neither is it meaningful, according to the holist position, to investigate an aspect or phenomenon of nature as if it were isolated from the rest of nature. The result of such an effort would not reflect the normal behavior of that aspect or phenomenon, since in reality it is not isolated at all, but is interrelated with and integrated in all of nature, including ourselves.

On the other hand lies the approach called *reductionism*, which holds that nature is indeed understandable as the sum of its parts. According to the reductionist position nature should be studied by analysis, should be 'chopped up' (mostly conceptually, of course) into simpler component parts that can be individually understood. (By 'parts' we do not necessarily mean actual material parts; the term might be used metaphorically. An example of that is presented in Sect. 2.2.3.) A successful analysis should then be followed by synthesis, whereby the understanding of the parts is used to help attain understanding of larger parts compounded of understood parts. If necessary, that should then be followed by further synthesis, by the consideration of even larger parts, comprising the understood compound parts, and attaining understanding of the former with the help of the understanding achieved thus far. And so on to the understanding of ever larger parts, until we hopefully reach an understanding of all of nature.

Now, each of the poles of holism and reductionism has a valid point to make. Nature is certainly interrelated and integrated, at least in principle, and we should not lose sight of that fact. But if we hold fast to extreme holism, everything will seem so fearsomely complicated that it is doubtful we will be able to do much science. Separating nature into parts seems to be the only way to search for simplicity within nature's complexity. Thus, the method of *reduction* shares with reproducibility and predictability their eminence as components of the foundation of science, and we should think of all three as forming the triple foundation of science.

Yet, an approach of extreme reductionism might also not allow much scientific progress, since nature might not be as amenable to reduction as such an approach claims, and the reductionist method of science might eventually run up against a holistic barrier. So science is forced to the pragmatic mode of operating *as if* reductionism were valid and adhering to that assumption for as long as it works. In the meanwhile science continues to operate very well. But it should be kept in mind that the inherent integrity of nature can raise its head at any time and indeed does so. The most well known aspect of nature's irreducibility is nature's quantum character [7,8].

Since all of nature, including ourselves, is interrelated and integrated, one might wonder how it is that reduction works at all. The answer lies in the choice of the part of nature that we apply reduction to. The coupling between that part and the rest of nature, including ourselves, must be sufficiently weak that it can be ignored or compensated for to a reasonably good approximation. In other words, the part of nature that is under investigation must be sufficiently isolated or isolable from the rest of nature, but not so much that we cannot observe or measure it. That consideration includes the well known issue of nondisruptive measurement: Our measurements must ideally not affect, and in practice only minimally affect, whatever it is that we are measuring. Thus the reductionist approach of science does not claim that the method of reduction is applicable in all cases. It does claim – and obviously correctly so – that nature offers us sufficiently many important situations in which the method of reduction works. But even when reduction is successful, it is a fact that the success is only an approximation, albeit possibly a very good one.

Reduction in science, the separation of nature into parts that can be individually understood, implies symmetry. The point is that if a reduction separates out a part that can be understood individually, then that part exhibits order and law (to be discussed in Sects. 2.2.3 and 2.4) and is explainable regardless and independently of what is going on in the rest of nature. In other words, the part of nature that is being individually understood possesses aspects that are immune to possible changes in the rest of nature. And this is symmetry:

1. *Possibility of a change.* It is possible to make many changes in the rest of nature.
2. *Immunity.* These changes do not affect important aspects of the part of nature that can be individually understood.

Reduction of nature can be carried out in many different ways. As the old saying goes, there's more than one way to slice a salami. We now consider three ways reduction is commonly applied in science, three ways nature is commonly 'sliced up', and we examine the symmetry implied by each.

2.2.1 Reduction to Observer and Observed

The most common way of reducing nature is to separate it into two parts: the *observer* – us – and the *observed* – the rest of nature. This reduction is so obvious, it is often overlooked. It is so obvious because in doing science we *must* observe nature to find out what is going on and what needs to be understood. Now, what is happening is this: Observation is interaction, so we and the rest of nature are in interaction, are interrelated, as was pointed out in the Sect. 2.2. Thus, anything we observe inherently involves ourselves too. The full phenomenon is thus at least as complicated as *Homo sapiens*. Every observation must include the reception of information by our senses, its transmission to our brain, its processing there, its becoming part of our awareness, its comprehension by our consciousness, etc. We appear to ourselves to be so frightfully complicated, that we should then renounce all hope of understanding anything at all.

So we reduce nature into us, on the one hand, and the rest of nature, on the other. The rest of nature, as complicated as it might be, is much less complicated than all of nature, since we have been taken out of the picture. We then concentrate on attempting to understand the rest of nature. (We also might, and indeed do, try to understand ourselves.

But that is another story.) However, as we saw above, since nature with us is not the same as nature without us, what right do we have to think that any understanding we achieve by our observations is at all relevant to what is going on in nature when we are not observing? The answer is that in principle we simply have no such right a priori. What we are doing is *assuming*, or adopting the working hypothesis, that the effect of our observations on what we observe is sufficiently weak or can be made so, that what we actually observe well reflects what would occur without our observation and that the understanding we reach under this assumption is relevant to the actual situation. This assumption might be a good one or it might not, its suitability possibly depending on the aspect of nature that is being investigated. It is ultimately assessed by its success or failure in allowing us to understand nature.

As is well known, the observer-observed analysis of nature is very successful in many realms of science. One example is Newton's explanation (Isaac Newton, English scientist and mathematician, 1642–1727) of Kepler's laws of planetary motion. That excellent understanding of an aspect of nature was achieved under the assumption that observation of the planets does not affect their motion substantially. In general, the reduction of nature to observer and observed seems to work very well from astronomical phenomena down through everyday-size phenomena and on down in size to microscopic phenomena. However, at the microscopic level, such as in the biological investigation of individual cells, extraordinary effort must be invested to achieve a good separation. The ever-present danger of the observation's distorting the observed phenomena, so that the observed behavior does not well reflect the behavior that would occur without observation, must be constantly circumvented.

At the molecular, atomic, and nuclear levels and at the subnuclear level, that of the so-called elementary particles and their structure, the observer-observed analysis of nature does not work. Here it is not merely a matter of lack of ingenuity or insufficient technical proficiency in designing devices that minimize the effect of the observation on the observed phenomena. Here it seems that the observer-observed interrelation cannot be disentangled *in principle*, that nature holistically absolutely forbids our separating ourselves from the rest of itself. Quantum theory successfully deals with such matters [7,8]. From it we learn that nature's observer-observed disentanglement veto is actually valid for *all* phenomena of *all* sizes. Nevertheless, the *amount* of residual observer-observed involvement, after all efforts have been made to separate, can be characterized more or less by something like atom size.

Thus an atom size discrepancy in the observation of a planet, a house, or even a cell is negligible, while such a discrepancy in the observation of an atom or an elementary particle is of cardinal significance.

One aspect of the symmetry implied by the observer-observed reduction, when this reduction is valid, is that the behavior of the rest of nature (i.e., nature without us) is unaffected by and independent of our observing and measuring. This behavior is thus an aspect of nature that is immune to certain possible changes, the changes being changes in our observational activities. It is just this symmetry that allows the compilation of objective, observer-independent data about nature that is a sine qua non for the very existence of science. It is intimately related to the symmetry that is reproducibility, which is discussed separately in Sect. 2.3.

Inversely, another aspect of this symmetry is that our observational activity is unaffected by and independent of the behavior of the rest of nature, at least in certain respects and to a certain degree. For example, if we had an ideal thermometer, we would apply exactly the same temperature measurement procedure regardless of the system whose temperature is being taken. (In practice, of course, things are not so simple and instead we make use of a consistent set of temperature measuring devices.) The symmetry here is that our observational activity is an aspect of nature that is immune to changes in what is being observed. This symmetry allows the setting up of measurement standards and thus allows the meaningful comparison of observational results for different systems. For instance, we can meaningfully compare the temperature of the sea with that of the atmosphere.

I might add that objectivity, too, is symmetry. Objective data are data that all observers agree about and whose validity is independent of observer. This means that the data are immune in some way to changes of observer. The symmetry is:

1. *Possibility of a change*. Observers can be changed.
2. *Immunity*. The validity of the data is independent of the observer.

That is certainly the general idea, although matters are not at all as simple as they might appear here. For the purposes of the present discussion we forgo the complications, some of which can be seen in [9].

Let me mention that there are fields of study, such as psychology, sociology, anthropology, and economics in which the observer-observed separation can be very difficult, if not altogether impossible, due to the strong interaction between the observer and the observed. Workers in such fields must take extra care to ensure that their data are in fact objective and observer-independent. If that is not achievable, then they are not doing scientific research.

2.2.2 Reduction to Quasi-Isolated System and Environment

Whenever we reduce nature into observer and the rest of nature, we achieve simplification of what is being observed, because instead of observing all of nature, we are then observing only what is left of nature after we ourselves are removed from the picture. Yet even the rest of nature is frightfully complicated. That might be overcome by further slicing of nature, by separating out from the rest of nature just that aspect or phenomenon that especially interests us. For example, in order to study liver cells we might remove a cell from a liver and examine it under a microscope.

But what right have we to think that by separating out a part of nature and confining our investigation to it, while completely ignoring the rest of nature, we will gain meaningful understanding? We have in principle no right at all a priori. Ignoring everything going on outside the object of our investigation will be meaningful if the object of our investigation is not affected by what is going on around it, so that it really does not matter what is going on around it. That will be the case if there is no interaction between it and the rest of nature, i.e., if the object of our investigation is an *isolated system*.

Now, an isolated system is an idealization. By its very definition we cannot interact with, thus we cannot observe, an isolated system, so no such thing can exist in nature, where nature is, we recall, the material universe with which we can, or can conceivably, interact. So we have no choice but to deal with nonisolated systems. Known anti-isolatory factors include the various forces of nature, which can either be effectively screened out or can be attenuated by spatial separation [10]. Additional known anti-isolatory factors involve quantum effects and inertia, which can be neither screened out nor attenuated. Thus even the most nearly perfectly isolated natural system is simply not isolated, and I therefore prefer the term 'quasi-isolated system' for a system that is as nearly isolated as possible. Or better, perhaps, a system that is sufficiently isolated for the purpose at hand. This is obviously not a clear-cut matter and involves approximation.

The separation of nature into *quasi-isolated system* and *environment* will be a reduction, if, in spite of the system's imperfect isolation, there are aspects of the system that are nevertheless unaffected by its environment, at least to a sufficient extent. And the fact of the matter is that the investigation of quasi-isolated systems does yield meaningful understanding, thus proving quasi-isolation to be a reduction of nature. Indeed, science successfully operates and progresses by the double reduction of nature into observer and observed and the observed into quasi-isolated system and its environment.

One side of the symmetry implied by this reduction is that those aspects of quasi-isolated systems that are not affected by their environment are aspects of nature that are immune to possible changes, the changes being changes in the situation of the environment. So the symmetry is:

1. *Possibility of a change.* Changes can be made in the environment of a quasi-isolated system.
2. *Immunity.* A quasi-isolated system possesses aspects that are not affected by certain such changes.

This symmetry is intimately related to the symmetry that is predictability, which will be discussed in Sect. 2.4. Inversely, due to the mutuality of interaction or of lack of interaction, there are also aspects of the environment of quasi-isolated systems that are immune to certain changes in the states of the quasi-isolated systems. This is another side of the symmetry implied by this reduction.

The reduction into quasi-isolated system and environment is not always possible. For complex systems whose components are in strong mutual interaction, such as, perhaps, social and economic systems, the reduction to significantly quasi-isolated relatively simple subsystem and environment (the rest of the complex system together with the rest of the world) might be difficult or impossible.

2.2.3 Reduction to Initial State and Evolution

The previous two ways of reducing nature – separation into observer and observed and separation into quasi-isolated system and its environment – are literal applications of the reductionist approach. The present way of reducing is a metaphoric application, or a broadening of the idea of a part of nature. Rather than a separation that can usually be envisioned spatially – observer here, observed there, or quasi-isolated system here, its environment around it – the present reduction is a conceptual separation, the separation of natural processes into *initial state* and *evolution*.

Things happen. Events occur. Changes take place. Nature evolves. That is the relentless march of time. The process of nature's evolution is of special interest to scientists, since predictability, one of the cornerstones of science, has to do with telling what will be in the future, what will evolve in time. Nature's evolution is certainly a complicated process. Yet order and law can be found in it, when it is properly sliced. First the observer should separate himself or herself from the rest of nature. Then he or she should narrow the scope of investigation from all of the rest of nature to quasi-isolated systems and investigate the natural evolution of such systems only. Actually, it is only for quasi-isolated systems that order and law are found.

Finally, and this is the present point, the natural evolution of quasi-isolated systems should be analyzed in the following manner. The evolution process of a system should be considered as a sequence of *states* in time, where a state is the condition of the system at any time. For example, the solar system evolves, as the planets revolve around the Sun and the moons revolve around their respective planets. (For simplicity we are ignoring other components of the solar system.) Now imagine that some duration of this evolution is recorded on a video cassette or a DVD. Such a recording is actually a sequence of still pictures. Each still picture can be considered to represent a state of the solar system, the positions of the planets and moons at some time. The full recording – the cassette or DVD – represents a segment of the evolution process. This illustration is deficient, however, since a state of the solar system really involves not only the positions of the planets and moons at some time, but also their velocities at the same time, while still pictures do not show velocities.

Then the state of the system at every time should be considered as an *initial state*, a precursor state, from which the following remainder of the sequence develops, from which the subsequent process evolves. For the solar system, for instance, the positions and velocities of the planets and moons at every single time, such as when it is twelve o'clock noon in Rockville on 20 October 2008, say, or any other time, should be considered as an initial state from which the subsequent evolution of the solar system follows.

When that is done, when natural evolution processes of quasi-isolated systems are viewed as sequences of states, where every state is considered as an initial state initiating the system's subsequent evolution, then it turns out to be possible to find order and law. (Order and law are discussed in more detail in Sect. 2.4.) What turns out is that, with a good choice of what is to be taken as a state for a quasi-isolated system, one can discover a law that, given *any* initial state, then gives the state that evolves from it at *any* subsequent time. Such a law, since it is specifically concerned with evolution, is referred to as a *law of evolution*.

For an example let us return to the solar system. It turns out that the specification of the positions of all the planets and moons at any single time is insufficient for the prediction of their positions at later times. Thus the specification of states solely in terms of position is not a good one for the purpose of finding lawful behavior. However, the description of states by both the positions and the velocities of the planets and moons at any single time does allow the prediction of the state evolving from any initial state at any subsequent time. The law of evolution in this case consists of Newton's three laws of motion and law of gravitation.

So the reduction needed to enable the discovery of order and law in the natural evolution of quasi-isolated systems is the conceptual splitting of the evolution process into initial state and evolution. The usefulness of such a separation depends on the independence of the two 'parts', on whether for a given system the same law of evolution is applicable equally to any initial state and whether initial states can be set up with no regard for what will subsequently evolve from them. Stated in other words, the analysis of the evolution process of a quasi-isolated system into initial state and evolution will be a reduction, if, on the one hand, nature indeed allows us, at least in principle, complete freedom in setting up the initial state (i.e., if nature is not at all concerned with initial states), while, on the other hand, what evolves from an initial state, once it is set up, is entirely beyond our control.

Let us consider this specifically for the example of the solar system and Newton's laws of motion and gravitation. The above reduction into initial state and law of evolution is successful because the two 'parts' are indeed independent. Newton's laws are indeed applicable to the solar system in whatever state it might be, no matter where the planets are located regardless of their velocities. On the other hand, there is nothing in Newton's laws that precludes us, at least in principle, from setting up a solar system in any state (of planetary positions and velocities) that we choose.

This reduction of evolution processes into initial state and evolution has proved to be admirably successful for everyday-size quasi-isolated systems and has served science faithfully for ages. Its extension to the very small seems quite satisfactory, although when quantum theory becomes relevant, the character of initial states becomes quite different from what we are familiar with in larger systems. Its extension to the large, where we cannot actually set up initial states, is also successful. But we run into trouble when we consider the Universe as a whole. One reason for this is that the concepts of order and law are scientifically meaningless for the evolution of Universe as a whole [6, 11]. Another reason is that it is not at all clear whether the concept of initial state is meaningful for the Universe. I do not think it is [12].

The symmetry that is implied by reduction into initial state and evolution follows immediately from the independence of the two 'parts', as described above. On the one hand, laws of evolution are an aspect of nature that is immune to possible changes, the changes being changes in initial states. On the other hand, initial states are an aspect of nature that is immune to possible changes, where the changes are hypothetical changes in laws of evolution, in the sense that initial states can be set up with no regard for what will subsequently evolve from them. So one symmetry of such a reduction is:

1. *Possibility of a change*. The state of a system, as an initial state, can be changed.
2. *Immunity*. The law of evolution for the system is the same no matter what its initial state.

While inversely and somewhat awkwardly:

1. *Possibility of a change*. The law of evolution of a system can be thought of as hypothetically varying.
2. *Immunity*. The system's state can be set up with no regard for what will evolve from it.

This symmetry, together with that implied by the reduction into quasi-isolated system and environment, is intimately related to the symmetry that is predictability, which is discussed in Sect. 2.4.

2.3 Reproducibility Is Symmetry

Science rests firmly on the triple foundation of reproducibility, predictability, and reduction. Science is concerned with the reproducible and predictable phenomena of nature, and any phenomenon that is either irreproducible or unpredictable or both, lies outside the domain of concern of science. Reproducibility is the replicability of experiments in the same lab and in other labs, which makes science a common human endeavor, rather than, say, a collection of private, incommensurate efforts. It makes science as much as possible an objective, or at least intersubjective, endeavor. (Intersubjectivity means that even if we are not sure about the objectivity of science, i.e., about its independence of any kind of observer, at least all *human* observers agree about what is going on. Concerning objectivity as symmetry, see [9].) We now show that reproducibility is symmetry, and we also show that reproducibility implies analogy (which, as we saw in Sect. 1.3, is symmetry).

Let us express matters in terms of experiments and their results. Reproducibility is then commonly defined by the statement that the same experiment always gives the same result. But what is the 'same' experiment? Actually each experiment, and we are including here even each run of the same experimental apparatus, is a unique phenomenon. No two experiments are identical. They must differ at least in time (the experiment being repeated in the same lab) or in location (the experiment being duplicated in another lab), and might, and in fact almost always do, differ in other aspects as well, such as in spatial orientation (since Earth revolves and rotates). So when we specify 'same' experiment and 'same' result, we actually mean equivalent in some sense rather than identical. We cannot even begin to think about reproducibility without permitting ourselves to overlook certain differences, where those differences involve time, location, and orientation, as well as various other aspects of experiments.

Consider the difference between two experiments as being expressed by the change that must be imposed on one experiment in order to make it into the other. Such a change might involve temporal displacement, if the experiments are performed at different times. It might (also) involve spatial displacement, if they are (also) performed at different locations. If the experimental setups have different directions in space, the change will involve rotation. If they are in different states of motion, a boost (a change of velocity) will be involved. We might bend the apparatus. We might replace a brass part with a plastic one. Or we might measure velocity rather than pressure. And so on.

But not all possible changes are changes we associate with reproducibility. Let us list those we do. We certainly want temporal displacement, to allow the experiment to be repeated in the same lab, and spatial displacement and rotation, to allow other labs to perform the experiment. The motion of Earth requires spatial displacement and rotation even for experiments performed in the same lab as well as velocity boosts for those performed at different times or locations. Then, to allow the use of different sets of apparatus, we need replacement by other materials, other atoms, other elementary particles, etc. Due to unavoidably limited experimental precision we must also include small changes in the conditions. And we also need changes in quantum phases (which are particularly quantum properties of systems, unrelated to such as 'liquid phase'), over which we have no control in principle. Those are the most apparent changes associated with reproducibility.

Let us denote the set of all changes we associate with reproducibility – and add any I might have overlooked – by REPRO. We now define reproducibility as follows: Consider an experiment and its result, consider the experiment obtained by changing the original one by any change belonging to REPRO, and consider the result obtained by changing the original result in the same way. If the changed result is what is actually obtained by performing the changed experiment, and if this relation holds for all changes belonging to REPRO, we have reproducibility.

As an example, imagine some experiment whose result is a violet flash emanating from some point in the apparatus some time interval after the switch is turned on. Now imagine repeating the experiment with the same apparatus, in the same direction and state of motion relative to Earth, etc., but 8 1/2 hours later and at a location 2.2 kilometers east of the original location. If a violet flash now appears 8 1/2 hours later than and 2.2 kilometers east of its previous appearance, we have evidence that the experiment might be reproducible. (As we know in this business, whereas a single negative result disproves reproducibility, no number of positive results can prove it. A few positive results make us suspect reproducibility; many will convince us; additional positive results will confirm our belief.)

Symmetry is materializing here; reproducibility is indeed symmetry. We see that in this way. Consider a reproducible experiment and its result. Change it and its result together by any change belonging to REPRO. The pair (changed experiment, changed result) is in general different from the pair (original experiment, original result), but there is an aspect of the pairs that is immune to the change. This aspect is the relation – call it physicality, actuality, reality, or whatever – that *the result is what is actually obtained by performing the experiment.* Said in other words, the symmetry that is reproducibility is that, for any reproducible experiment and its result, the experiment and result derived from them by any change belonging to REPRO are also an experiment and its actual result. Expressed in our two-point schema, the symmetry is:

1. *Possibility of a change*. Changes belonging to REPRO can be performed on the experiment and its result.
2. *Immunity*. The changed result remains the actual result of the changed experiment.

Reproducibility implies analogy, discussed in Sect. 1.3. The analogy is that the changed experiment is to the changed result as the original experiment is to the original result for all changes belonging to RE-PRO, with the relation that the result in each case is what is actually obtained by performing the experiment.

Earlier in this chapter, in Sect. 2.2.1, I stated that the symmetry implied by the observer-observed reduction is intimately related to the symmetry that is reproducibility. Both symmetries involve immunity under changes in observational activities. And in both cases the immunity is that the observed behavior does not change. So reproducibility, as a component of the foundation of science, is not independent of one of the other components, reduction.

2.4 Predictability Is Symmetry

The other constituent of the foundation of science, along with reproducibility and reduction, is predictability. Predictability means that among the phenomena investigated, order can be found, from which laws can be formulated, predicting the results of new experiments. Then theories can be developed to explain the laws. We now show that predictability, too, is symmetry and show as well the analogy that predictability implies.

Here again we express things in terms of experiments and their results. Predictability, then, is that it is possible to predict the results of new experiments. Of course, that does not often come about through pure inspiration, but is much more usually attained by performing experiments, studying their results, finding order, and formulating laws.

So imagine we have an experimental setup and run a series of n experiments on it, with experimental inputs $\exp_1, \ldots, \exp_n$, respectively, and corresponding experimental results $\text{res}_1, \ldots, \text{res}_n$. We then study those data, apply experience, insight, and intuition, perhaps plot them in various ways, and discover order among them. Suppose we find that all the data obey a certain relation – denote it R – such that all the results are related to their respective inputs in the same way. Using function notation, we find that $\text{res}_k = R(\text{exp}_k)$, $k = 1, \ldots, n$. This relation is a candidate for a law $\text{res} = R(\text{exp})$ predicting the result res for *any* experimental input exp. Imagine further that this is indeed the correct law. Then additional experiments will confirm it, and we will find that $\text{res}_k = R(\text{exp}_k)$ also for $k = n + 1, \ldots$, as predicted. Predictability is the existence of such relations for experiments and their results.

For an example, consider again a given sphere rolling down a fixed inclined plane, with the experimental procedure of releasing the sphere from rest, letting it roll for any time interval t, and noting the distance d the sphere rolls in that time. Here t and d are playing the roles of exp and res, respectively. Suppose we perform ten experiments, giving the data pairs $(t_1, d_1), \ldots, (t_{10}, d_{10})$. We study the data and plot them in various ways. The plot of distance d_k against square of time interval t_k^2 looks like all ten points tend to fall on a straight line. That suggests the relation that the distance traveled from rest is proportional to the square of the time interval, $d_k = bt_k^2$, $k = 1, \ldots, 10$. And that in turn suggests the law $d = bt^2$ predicting the distance d for *any* time interval t. As it happens, this hypothesis is correct, and all additional experiments confirm it. The relation $d_k = bt_k^2$ is found to hold also for $k = 11, 12, \ldots$, i.e., also for data pairs $(t_{11}, d_{11}), (t_{12}, d_{12}), \ldots$. Thus, the relation of distance to time interval is a predictable aspect of the setup.

That predictability is symmetry can be seen as follows. For a given predictable experimental setup consider all the experiment–result pairs (exp, res) that have been, will be, or could be obtained by performing the experiment. Change any one of these into any other simply by replacing it. The changed pair is different from the original one in general, but the pairs possess an aspect that is immune to the change. This aspect is that *exp and res obey the same relation for all pairs*, namely, the relation $\text{res} = R(\text{exp})$. Put in different words, the symmetry is that for any experiment and its result, the experiment and its result obtained by changing the experimental input obey the same relation as the original experiment and result. In the two-point schema:

1. *Possibility of a change.* The input to the experiment can be changed.
2. *Immunity.* The experimental result maintains the same relation with the experimental input.

Just as for reproducibility, predictability implies analogy: For a predictable experimental setup any experiment is to its result as any other experiment is to its result. The same relation res = R(exp) holds for the input exp and result res in each case.

It might prove enlightening to consider a case study here. Let us study the archetypal case of Kepler and his three laws of planetary motion. While pondering the many and various astronomical phenomena known to him, Kepler found a certain order within the general confusion. He found a classification of the motions of the celestial bodies, whereby the motions of the then known planets were assigned to one class and the motions of all the other bodies were assigned to another. The significant characterization of the former class, that of the planets, was that in a heliocentric reference frame (1) the orbit of each body is an ellipse with the Sun at one of its foci, (2) the radius vector of each body from the Sun sweeps out equal areas in equal time intervals, (3) the squared ratio of the orbital periods of any two bodies equals the cubed ratio of their respective orbital major axes. The class of all the non-planets was characterized, of course, by the nonfulfillment of (1)– (3). Thus the planetary motions were transformed, through the order Kepler perceived, from a set of individual motions, each requiring its own explanation, to a class of motions requiring a common explanation.

As we saw in Sect. 1.3, classification implies and is implied by analogy. Thus the motions of the known planets were found to be analogous in that they all fulfilled (1)–(3), while the motions of all the other celestial bodies were analogous in that they did not fulfill (1)–(3). We also saw that analogy and classification are symmetry, and we just saw that this symmetry in the present case is the symmetry that is predictability. Kepler's order = classification = analogy suggested that (1)–(3) might be laws of planetary motion for all solar planets, not just for the then known ones, and indeed this hypothesis has been continually confirmed as additional planets have been discovered. To put things in terms of experiments and their results, the discovery of additional planets might be thought of as additional experiments, whose results fit the laws derived from the order perceived in the motions of the previously known planets, which might be thought of as previous experimental results. Kepler's laws inspired Newton to develop an explanatory theory in the form of his three laws of motion and his law of gravitation.

For another example of order, leading to law and predictability, leading to theory, we can consider the periodic table of the chemical elements that exemplified Ernest Rutherford's 'butterfly collecting' in the Preface and was discussed more thoroughly in Sect. 1.3. The order and law that the chemists found, as expressed by the periodic table, did indeed allow them to predict the existence of elements that were missing from the table and, moreover, predict their properties. When those elements, such as gallium and germanium, were eventually discovered, their properties were found to be very close to what was predicted. The explanation of the periodic table was eventually provided by physics, Rutherford's 'real science', in the form of quantum theory.

In Sect. 2.2.3 earlier in this chapter, I stated that the symmetry implied by reduction into initial state and evolution and that implied by reduction into quasi-isolated system and environment are intimately related to the symmetry that is predictability. Well, now is the time to justify the statement. Predictability is the result of order and law, in particular law of evolution. We can predict what the outcome of a situation will be, if we know the law of evolution for the system under consideration. This is reduction into initial state and evolution. No such reduction, no predictability. Hence they are related intimately and so, therefore, are the symmetries that they imply.

The possibility of usefully performing a reduction into initial state and evolution hinges on a successful reduction into quasi-isolated system and environment. Only then is the system sufficiently free of the uncontrollable and unpredictable influences of the environment to be able to exhibit orderliness and lawful behavior, which allow the possibility of a useful reduction into initial state and evolution and thus allow predictability. That intimately relates predictability and reduction into quasi-isolated system and environment and, accordingly, relates their implied symmetries. Q.E.D.

2.5 Analogy in Science

Since both reproducibility and predictability imply analogy and analogy is symmetry, I would be remiss if I did not take this opportunity to elaborate a bit on the role of analogy in science. It has been stated [13]: "The value of [...] analogies in stimulating research [... is] self-evident ...," and "... it cannot be denied that analogy plays an important role in scientific creativity." I would put it much more strongly and state that analogy is *essential* in *any* science-related activity. My justification for this claim is the following.

We showed in Sects. 2.3 and 2.4 that analogy lies at the foundation of science, specifically that both reproducibility and predictability imply analogy. For reproducibility, any experiment-result pair can be changed by any of the set of changes associated with reproducibility, and the changed result is what is actually obtained by performing the changed experiment. Thus, all experiment-result pairs related by reproducibility-associated changes are analogous in that they all obey the same relation, namely, that the result is what is actually obtained by performing the experiment. For predictability, all actual and potential experiment-result pairs for the same predictable experimental setup are analogous in that they all obey the same law, usually expressed as a mathematical relation, as in the example of the rolling sphere.

Zooming in on predictability and order, the analogy involved there is more familiar than might be thought. After all, order, in any sense of the term, involves classification, which implies analogy. Thus, however one prefers to look at the matter, analogy is essential for any science related activity, be it science teaching, science research, or scientific creativity.

For discussions of analogy in science see [14], for a very general discussion, and [13], where analogy symmetry is given the name 'logical symmetry' and its pertinence to classification in science and to the periodic table of the chemical elements in particular is discussed.

Examples are certainly warranted here, and we might start with Kepler's three laws of planetary motion, which served us so well in Sect. 2.4. Kepler concluded from the astronomical data available to him that the motions of all the observed planets obeyed the same laws. That made the planets analogous, introduced order into the astronomers' picture of the solar system, and enabled Newton to derive his three laws of motion and law of gravitation.

For another example, Mendeleev's periodic table of the chemical elements exhibited order and analogies among the elements [13]. Those analogies were a major driving force in the development of models of atomic structure and in the development of quantum theory to explain the structure.

For yet another example, the analogies discovered among the elementary particles [10,15,16] were and are essential to the development of theories of their behavior and structure.

2.6 Symmetry at the Foundation of Science

We are now in a position to recognize and appreciate how large a role symmetry plays in the foundation of science. First, reproducibility, a major component of that foundation, is symmetry. And reproducibility implies analogy, which is symmetry also (see Sect. 2.3). Second, predictability, another major component of the foundation of science, is symmetry too. And it, too, implies analogy, which again is symmetry (refer to Sect. 2.4). And third, reduction, which forms the third major component of the foundation of science, is symmetry (see Sect. 2.2). In addition, objectivity is symmetry (see Sect. 2.2.1).

What is going on here? Where does all this symmetry come from? What does it mean? What is this telling us?

It seems to me that to answer these questions we should recall that science, whose foundation, as we just learned, is flooded with symmetry, is *our attempt to understand rationally and objectively the reproducible and predictable aspects of nature.* Note: (1) *Our* attempt. (2) 'Understanding' here refers to *our* understanding. (3) Reproducibility has to do with *our* investigational activities. (4) Predictability means that *we* can predict. So science is very much about *us.* Of course, for science to succeed, nature must possess reproducible and predictable aspects and be understandable rationally and objectively. So the success of science is saying something about nature itself, about the ontology of nature.

But the 'we/us/our' character of science is an epistemological matter, having to do with the acquisition of knowledge about nature and achievement of understanding. So the abundance of symmetry at the foundation of science reflects on the way we view nature: We see nature through symmetry spectacles. Apparently it is built into our innate perceptional makeup to pay special attention to patterns and to notice order, analogy, and symmetry. Presumably that trait evolved in *Homo sapiens* as an advantageous one, since it does prove to be very useful.

Might we then go so far as to claim that science *is* symmetry? Well, I suppose this is a matter of taste. I would indeed go so far. But even if you disagree with me about this, I hope you do agree that it is not too much of a stretch.

We will see in Chap. 3 that science reveals many and diverse symmetries in nature. Indeed, it turns out that our understanding of what appear to be very fundamental aspects of nature is couched in terms of symmetry. So not only do we *view* nature through symmetry spectacles, but we *understand* nature in the language of symmetry. Yet it is hardly justified to claim that nature *is* symmetry, at least at our present stage of understanding. We understand enough to know that there is much more to understand. It is clear that deeper levels of nature must underlie what now seems to be fundamental. The way nature has been revealing itself through science so far would lead us to expect that symmetry will be found to play a major role at those very fundamental levels. Perhaps it will be found that in some sense nature indeed *is* symmetry. That is something to look forward to. But more about this in Chap. 3.

2.7 Summary

We started out in Sect. 2.1 by defining nature as the material universe with which we can, or can conceivably, interact. Science, then, is our attempt to understand rationally and objectively the reproducible and predictable aspects of nature. In Sect. 2.2 we saw that science operates by the method of reduction, by 'slicing' the Universe into 'parts', and that reduction implies symmetry. Three ways that reduction is applied in science are: observer and observed, quasi-isolated system and environment, and initial state and evolution, discussed in Sects. 2.2.1, 2.2.2, and 2.2.3, respectively. The symmetry implied by each reduction was pointed out. It is for quasi-isolated systems that order and law are found, so for such systems reduction into initial state and evolution can be useful.

Science rests firmly on the triple foundation of reproducibility, predictability, and reduction. In Sect. 2.3 we discussed in detail reproducibility, the possibility of replicating experiments, and saw how reproducibility is symmetry and how it implies analogy, which is symmetry too. Then, with a detailed discussion in Sect. 2.4, we saw the same for predictability, the existence of order from which laws can be formulated, predicting the results of new experiments. In Sect. 2.5 we elaborated a bit on the role of analogy, which is symmetry, in science and in any science related activity: It is absolutely essential. And putting it all together, we saw in Sect. 2.6 that via reproducibility, predictability, and reduction, and with the help of analogy, symmetry holds major importance in the foundation of science. So much so, in fact, that one might well claim that science *is* symmetry.

3

Symmetry in Physics

Science possesses a hierarchical structure, in which each of its various hierarchy levels, or 'sciences', maintains a rather independent existence. The science of chemistry, for instance, can operate largely within its own framework, without reference to other sciences. Similarly for biology, physics, psychology, and sociology, as examples. However, when explanations are required from outside any particular science, such explanations have very well defined directions, which impose order in the hierarchy of the sciences.

Extrachemical explanations about chemical phenomena come only from physics and from no other science. When an explanation for a biological effect is required from outside biology, its source invariably lies either in chemistry or in physics, or possibly in both, and not in any other science. The science of mind, psychology, is more and more obtaining extrapsychological explanations from biology. Extrasociological explanations for sociological phenomena come from psychology. Physics, alone among the sciences, has no need for any of the other sciences to explain anything within its purview.

Thus a hierarchy exists, in which physics can be pictured at the bottom, in foundation position, chemistry just above physics, biology lying above chemistry, psychology above biology, and sociology on top of psychology. Every level obtains explanations solely from lower levels. Thus, all explanations from one science to another eventually reach physics and physics is therefore the most fundamental science.

As the most fundamental of all sciences, physics studies the most fundamental aspects of nature. It turns out that those aspects reveal much symmetry, even beyond the symmetries that allow science to operate through reproducibility, predictability, and reduction, with the help of analogy and objectivity, as described in Chap. 2. In the present chapter we start discussing how physics deals with the symmetries of nature and what kinds of symmetry nature presents.

In line with the somewhat more technical character of this chapter compared with the previous ones, our terminology becomes more technical as well. We start by replacing the term 'change', in the sense of what symmetry is immunity under, with *transformation*, which is the conventional usage.

3.1 Symmetry of Evolution

Two important and useful manifestations of symmetry in physics are symmetry of evolution of quasi-isolated physical systems and symmetry of states of physical systems. We start with a discussion of the former.

Physical systems evolve. Systems evolve from initial states to final states, if considered discretely, or considered continuously, the states of systems are functions of time. Physics is concerned with the evolution of quasi-isolated systems, because their evolution is found to have reproducible and predictable features (see Sect. 2.2.2). Indeed, the choice of what is to be considered a state for a system is made in such a way as to maximize those features.

(Note the circularity here. A physical system is declared quasi-isolated if it exhibits reproducibility and predictability with suitable choice of state. In fact, it is by the lack of reproducibility and predictability that new effects and interactions are discovered, such as occurred in the discovery of the neutrino, for example.)

Symmetry of evolution of quasi-isolated systems means that there is some transformation that, if applied to any physically possible evolution, would result in another physically possible evolution, and if hypothetically applied to any impossible evolution, would result in another impossible evolution. For quantum phenomena it would result in an evolution having the same probability. The aspect of evolutions that is immune to the transformation is then their possibility, impossibility, or probability. Such symmetry is reflected in the laws and theories describing and explaining the evolution of quasi-isolated systems and as such is also called *symmetry of the laws of nature*. Symmetries of evolution of quasi-isolated systems, or symmetries of the laws of nature, are intimately associated with *conservations* (also called *conservation laws*), such as conservation of energy, conservation of momentum, and conservation of electric charge. We elaborate on that in Sect. 3.7.

As in the discussion of reproducibility in Sect. 2.3, in the present discussion, too, it is convenient to express things in terms of experiments (initial states of quasi-isolated systems) and their results (the final states evolving from them). So consider some transformation applied both to an experiment and to its result. If the actual result of performing the transformed experiment is the same as the transformed result of the original experiment and if that is valid for all experiments, we will have symmetry of evolution (or, symmetry of the laws of nature). In other words, we will have symmetry of evolution under a transformation, if for any experiment and its result, the experiment and result derived from them by the transformation are also an experiment and its actual result. The aspect of experiment-result pairs that is immune to the transformation is, just as for reproducibility, that *the result is what is actually obtained by performing the experiment.*

The physical significance of symmetry of evolution is that nature is indifferent to certain aspects of physical systems, that the evolution of physical systems is independent of certain of their aspects. Or stated in other words, certain aspects of physical systems are irrelevant to the systems' evolution. All transformations affecting only those aspects of physical systems that are irrelevant in this sense are symmetry transformations of evolution, and any transformation affecting a relevant aspect cannot be a symmetry transformation of evolution. So two systems differing only in irrelevant aspects will evolve in exactly the same way save for their (irrelevant) difference, and thus their difference will be preserved throughout their evolution, resulting in final states that differ precisely and solely as did the respective initial states. The transformation bringing about this difference is thus a symmetry transformation of evolution.

When symmetry of evolution is reflected in laws and theories as symmetry of the laws of nature, it is sometimes referred to as expressing 'impotence'. The idea is that the laws and theories are 'powerless' to grasp and take into consideration the irrelevant aspects involved in the symmetry, those aspects to which nature is indifferent. Thus indifference of nature is exhibited as impotence of laws and theories.

Now, nature does indeed possess such symmetries. As far as is known at present, the universal symmetries of evolution of quasi-isolated systems are symmetries under spatial displacements (changes of location), temporal displacements (changes of time), rotations (changes of orientation), changes of velocity (also called boosts) up to the speed of light [or, taken together and formulated group theoretically, symmetry under the Poincaré group of space-time transformations (Jules Henri Poincaré, French mathematician, 1854–1912), the symmetry required by the special theory of relativity], and a certain kind of change of phase (a more abstract kind of transformation) denoted U(1). (Here the term 'phase' refers to a quantum characteristic of a system and has nothing to do with such as 'solid phase'. Note that many of these symmetries are involved in the symmetry that we already know as reproducibility, which was discussed in Sect. 2.3, so that the very existence of science already implies certain symmetries of evolution, but we refrain from elaborating on the point.) Expressed in another way, the just-mentioned universal symmetries mean that with regard to the evolution of quasi-isolated systems nature does not recognize, respectively, absolute position, absolute instant, absolute direction, absolute velocity, or absolute phase.

As an example, symmetry of evolution under spatial displacements, the indifference of nature to position, can be expressed in this manner: Any two physical systems that are simultaneously in identical states, except for one being here and the other being there, will evolve into final states that are also simultaneously identical, except for one being here and the other there, respectively, for all heres and theres. Thus, nature does not recognize absolute position through the evolution of quasi-isolated systems. Laws and theories must accordingly be impotent with regard to position, and the only position variables that can be allowed to enter them are relative positions. So if one is developing a mathematical theory of, say, the interaction of two particles, the only position variable that can appear in the theory is the relative position of one particle with respect to the other.

Or, we can express the symmetry of evolution under boosts (velocity changes), the indifference of nature to velocity, this way: Any two physical systems that are simultaneously in identical states, except for one moving at constant velocity with respect to the other, will evolve into final states that are also simultaneously identical, except for one moving at the same constant velocity with respect to the other, respectively, for all velocities up to the speed of light. In that way nature does not recognize absolute velocity through the evolution of quasi-isolated systems. Accordingly, laws and theories must be impotent with regard to velocity, and the only velocity variables allowed in them are relative velocities.

As another example of symmetry of evolution, but not a universal symmetry, consider any macroscopic system, which can be described both in terms of its macrostates and in terms of its microstates, where every macrostate represents a class of corresponding microstates. The macroscopic evolution of such a system is indifferent to the actual microstate realizing its macrostate, or in other words, its microstate is an irrelevant aspect of its macrostate with regard to its macroevolution. Thus we have symmetry of macroevolution under the transformation of permuting, i.e., switching around, microstates corresponding to the same macrostate. This symmetry is not universal, because not all systems are macroscopic and have the dual characterization in terms of microstates and macrostates.

To be more specific, imagine a quantity of pure ideal gas in thermal contact with a heat bath, which is a system whose job is to maintain a constant temperature. The gas's macrostates can be specified by the quantities: pressure, volume, temperature, and amount of gas (for instance, the number of molecules or moles). The macroevolution of the gas is described by the ideal gas law $pV = nRT$, where p denotes the pressure, V the volume, n the quantity of gas, T the absolute temperature, and R is a suitable constant. The gas's microstates can be specified by the position and velocity of every one of its molecules. Its microevolution is described by Newton's laws of motion and whatever intermolecular force law applies. The heat bath maintains a constant temperature, by definition, and we need not get involved with its microstates.

Now, imagine starting with the gas in some macrostate and the heat bath at twice the gas's absolute temperature and putting the two in contact, while maintaining a constant quantity of gas at a constant volume. Then the system will very rapidly evolve to equalize temperatures and thus evolve to a macrostate in which the gas has twice the absolute temperature and twice the pressure than it did before the process and in which the bath temperature is unchanged. That evolution is independent of the particular microstates the gas and the bath are in at any time. Whatever the combined system's initial microstate, it will evolve to some microstate among those that correspond to its final macrostate.

It is also a possibility, and in fact it happens, that nature possesses symmetries of evolution that are not valid for all systems, but only for certain ones of them. Such inexact symmetries are described, according to the case, as 'partial', 'limited', or 'approximate'. For example, symmetry of evolution under the transformation of particle-antiparticle conjugation (the replacement of every particle with its corresponding antiparticle) is valid only for all systems that do not involve neutrinos. Also, symmetry of evolution under the transformation of spatial inversion (reversal of all directions, which is equivalent to mirror reflection combined with a 180° rotation) is valid only for systems that do not involve neutrinos. Yet symmetry of evolution under the combined transformation of particle-antiparticle conjugation and spatial inversion *is* valid for systems involving neutrinos, but not for those that involve neutral kaons, for instance.

Consider also the transformation of time reversal. This has nothing to do with time running backwards, whatever that might mean. It is the replacing of an evolution with an evolution that starts from the end of the original evolution and ends at the original evolution's beginning. Think of it as running a video, DVD, or movie in reverse. Symmetry of evolution under time reversal means that the original evolution and its time reversed counterpart would either both be physically possible, both be physically impossible, or both have the same probability. This symmetry is not valid for macroscopic systems. Whereas a dropped raw egg splatters and makes a mess, such a mess is never found to collect itself together and jump up into one's hand as an intact egg. Yet at the level of the fundamental interactions, symmetry of evolution under time reversal *is* valid in general, but not for a class of systems including those that involve neutral kaons.

Earlier in this section we mentioned change of phase. One universal symmetry of evolution of quasi-isolated systems is indeed symmetry under a certain kind of phase change, denoted U(1). Other kinds of phase change are associated with partial symmetry of evolution. Evolutions that are governed by the strong interaction possess symmetry under a certain kind, denoted SU(3), while this symmetry is invalid for evolutions controlled by the weak interaction. The latter evolutions have their own symmetry under a different kind of phase change, whose symbol is SU(2).

3.2 Symmetry of States

Another important and useful manifestation of symmetry in physics is the symmetry of states of physical systems. This manifestation would seem reasonably straightforward: We have symmetry of states, or a state is symmetric, if it is possible to change a state in a way that leaves some aspect of it intact. That, however, is opening a Pandora's box of triviality and boredom, since very many physical systems are sufficiently complex that there are very many physically trivial and uninteresting transformations that leave intact very many physically trivial and uninteresting aspects of their states.

Well, if it is physical significance and interest we want, and that is certainly what scientists should want, we had better let nature guide us. And nature has spoken: Let the immune aspects of states be their irrelevant aspects in the sense of Sect. 3.1, those aspects to which nature is indifferent and with regard to which the laws and theories are impotent, whether totally, partially, approximately, or to some limited degree. The transformations under which states are symmetric, then, are the transformations involved in the symmetries of evolution, again either exact symmetries or partial, approximate, or limited symmetries. Under those transformations *the transformed state is indistinguishable (or approximately indistinguishable) by nature from the original*.

That being the situation, then how, one very well might ask, do the transformed and original states differ at all, and just what transformations were actually performed? Indeed, there is no difference between the transformed and original states and, indeed, the transformations are invisible – within the context of the quasi-isolated system for whose evolution the immune aspects of states are irrelevant, so that the state is symmetric under the transformations. That purely and simply follows from our definitions. The states are distinguishable and the transformations detectable only with respect to some suitable external system to which the transformations are not applied. Such a system is a reference frame for the transformations. (We discuss reference frames in Sect. 3.3.)

As an example, consider the action of any of the universal symmetry transformations mentioned in Sect. 3.1, say spatial displacement, on any state of a quasi-isolated system. Within the system the transformed and original states do not differ in any way that nature can distinguish; their difference is irrelevant. Put in more familiar terms, no experiment carried out wholly within the system can detect any difference between the states. Or in other words, absolute position is undetectable within a quasi-isolated system. (However, they do differ with respect to a fixed coordinate system external to the system under consideration.)

For another example, replace 'spatial displacement' with 'boost' (change of velocity) in the previous example. The bottom line then becomes: Absolute velocity is undetectable within a quasi-isolated system. And so on for the other universal symmetry transformations: rotations and certain phase changes and the undetectability within quasi-isolated systems of absolute direction and absolute phase. And similarly for the limited symmetry transformations.

As an additional example, consider an equilaterally triangular homogeneous flat metal plate. The system possesses symmetry under rotations by 120° and 240° about its center within its plane with respect to appearance and macroscopic physical properties. This means that any triplet of states mutually related by those rotations are indistinguishable by means of external appearance and macroscopic physical properties. External appearance involves the evolution of light waves impinging on and absorbed and reflected by the surface of the plate, which means that such triplets of states absorb and reflect light in the same way, as far as our visual perception is concerned. That is the basic immunity that underlies unchanged external appearance. As for macroscopic physical properties, the states of such a triplet are actually microstates corresponding to the same macrostate, and the macroevolution involved in macroscopic physical properties cannot distinguish among them. That is the basic immunity here.

For another example, note that with respect to macroevolution, every macrostate of a quasi-isolated macroscopic system is symmetric under change of microstate realizing that macrostate. The difference between microstates corresponding to the same macrostate is irrelevant to the macroevolution of the system. For instance, every macrostate of a gas (specified, say, by pressure, volume, temperature, and quantity of gas) can be realized by a very large of number of microstates (characterized by the position and velocity of every molecule). As far as quasi-isolated macroevolution of the gas is concerned, it is immaterial which of all those microstates realizes a macrostate. If the gas evolves from some initial macrostate to some final macrostate, then whatever its initial microstate realizing its initial macrostate, its microevolution will take it into some final microstate realizing its final macrostate.

One of the most impressive – to me, at least – manifestations of state symmetry in nature is the crystalline state. The transformations under which every crystal is symmetric are spatial displacements by integer multiples of certain minimal displacements, which are the dimensions of the crystal's unit cell. (The unit cell of a crystal is the 'building block' of the crystal, in the sense that a perfect crystal can be viewed as constructed of very many replicas of the unit cell arrayed in precise spatial order in three dimensions.) In addition, a crystal might also possess symmetry under various rotations, reflections, and combinations of those. In actuality all these symmetries are approximate, due to the finite size of a crystal and to naturally occurring defects in the crystal structure. So when we think of crystalline symmetry, we are really thinking of a perfectly structured infinite crystal, called a crystal lattice.

The coexistence of rotation, reflection, and discrete displacement symmetries for a crystal lattice imposes severe limitations on the possible combinations of those symmetries, which are the subject of crystallography. For instance, it is rather easy to show that any axis of rotation symmetry of a crystal lattice must be either two-fold (i.e., symmetry under 180° rotations about the axis), three-fold (symmetry under 120° and 240° rotations), four-fold (symmetry under 90°, 180°, and 270° rotations), or six-fold (symmetry under rotations by integer multiples of 60°). Five-fold, seven-fold, and higher-fold rotation symmetry axes are simply incompatible with symmetry under spatial displacements.

Thus, it came as a very big surprise when five-fold rotation symmetry was experimentally detected in certain crystals [17]. It turned out, however, that the five-fold symmetry was not symmetry of the lattice as a whole – as indeed it could not be – but rather the symmetry of certain local, limited configurations of atoms (or ions or molecules, as the case may be) that happened to legitimately form part of the crystal structure.

Another very interesting and important phenomenon involving state symmetry has to do with phase transitions, such as the freezing of liquid water and its inverse, the melting of ice. (Phase transition has nothing to do with change of phase that we discussed in Sect. 3.1. 'Phase' in 'phase transition' refers to a macroscopic state of matter, such as solid, liquid, or gas, while 'phase' in 'change of phase' denotes an abstract, mathematical, quantum property of a system.) Let us follow just such a phase transition from the point of view of state symmetry. Imagine starting with liquid water. At any instant the water molecules and ions are distributed in quite a random manner throughout the liquid's volume. So at any instant the liquid does not possess any displacement, rotation, or reflection symmetry. However, the microscopic constituents of the liquid are in constant random motion. So on average over time the liquid can be considered to be homogeneous and isotropic and to be symmetric under all displacements, rotations, and reflections. Actually, as for a crystal, it is only an infinite volume of liquid that ideally possesses those symmetries. But we ascribe them to the actual volume of liquid.

When liquid water is cooled to its freezing point, 0°C under standard conditions, further removal of heat causes it to undergo a phase transition and become ice, a crystalline solid. There is also a concomitant reduction of symmetry. Instead of the liquid's symmetry under all spatial displacements, the solid possesses symmetry only under displacements that are integral multiples of certain minimal displacements. Instead of symmetry under all rotations about all axes, ice is symmetric only under certain rotations about certain axes. And instead of symmetry under all reflections, we find symmetry only under certain reflections. The phase transition creates distinctions among locations, directions, and orientations that were initially indistinguishable.

On the other hand, a solid-to-liquid phase transition brings about an increase of symmetry. The limited symmetry of the crystal state becomes the wider symmetry of a liquid. Initially distinct locations, directions, and orientations become indistinguishable.

A phase transition that is somewhat similar to the liquid-to-solid one occurs when a ferromagnetic material is cooled. This kind of material is one that exhibits a strong response to an applied magnetic field by becoming a magnet itself and greatly reinforcing the applied field. At sufficiently high temperature and in the absence of an external magnetic field, a ferromagnetic material is not magnetized. In a very sketchy description, the atoms of a ferromagnetic material are themselves tiny magnets that affect their neighbors strongly, so that there exists a tendency for the atoms to align with each other and reinforce each other's magnetic field. An external, applied magnetic field tends to force those elementary magnets to align with it. On the other hand, when the material is sufficiently warm, the random thermal motion of the atoms overcomes their tendency to align and, in the absence of an applied magnetic field, they take no distinguished direction on the average. So the material then possesses symmetry under all rotations about all axes.

As such a material is cooled, it reaches a temperature at which it undergoes a phase transition. In that transition the tendency of the atoms to align with each other overcomes the randomizing effect of their thermal motion, and the elementary atomic magnets align with and magnetically reinforce each other. The material spontaneously becomes a macroscopic magnet. That introduces a distinguished direction, the direction of the magnet's axis, which the material did not posses at higher temperatures. Thus the phase transition of spontaneous magnetization is accompanied by a reduction of symmetry. Symmetry under all rotations about all axes reduces to symmetry under all rotations about all axes in but a single direction.

For some other kinds of phase transition, see, for example, [18] and [19].

3.3 Reference Frame

In Sect. 1.2 we saw that for a transformation to be physically significant, a reference frame is needed. The reference frame must be affected by the transformation. So if a transformation is one that is involved in symmetry, an appropriate reference frame for the transformation must be asymmetric under the transformation. In the present section we discuss the idea of reference frame in detail.

A reference frame is a standard by which transformations are defined, performed, and detected and by which states are distinguished. By its very raison d'être *a reference frame cannot be immune to the transformations for which it is to serve as reference, nor can those transformations be applied to it.* Therefore, *a reference frame must be asymmetric under the transformations for which it is a reference; even if transformations are involved in symmetries of any kind, their reference frame must violate those symmetries.*

Accordingly, a reference frame for spatial displacement, for example, must possess a distinguished origin, must have scales marked off in three independent directions, and must be declared fixed in space and thus unaffected by any spatial displacing that might be occurring. Such a reference frame is indeed asymmetric under spatial displacement.

Or, for particle-antiparticle conjugation, which is the replacement of every particle with its corresponding antiparticle, an appropriate reference frame would be a set of standard declared particles – a proton, a neutron, an electron, etc. – preserved at an intergalactic bureau of standards, with which by comparison one could distinguish between proton and antiproton, neutron and antineutron, etc. Fortunately our own bodies furnish such a standard, at least for the proton, neutron, and electron. If the universe consisted solely of photons and gravitons, which are the respective antiparticles of themselves, there would be no reference frame for standard particles, the concept of particle-antiparticle would be meaningless, and so would particle-antiparticle conjugation and symmetry under it.

Transformations made on physical systems, where the transformations are referred to fixed reference frames, are called *active transformations*. That is the only kind of transformation we have been considering so far in this book. However, one can do things differently. In contrast, transformations can be made only on reference frames without affecting the physical systems under investigation. Such transformations are called *passive transformations*. Thus, while active transformations actually transform states to physically different states, passive transformations do not transform states but transform the descriptions of the same states, i.e., they transform only the names of states. A given transformation, with no specification of what it is supposed to affect, can freely be viewed either actively or passively, i.e., either as a transformation of state or as a transformation of reference frame, where the latter transformation is the inverse of the former. In other words, an active transformation can be given a passive interpretation and vice versa.

For example, a particle located at a point with x, y, z coordinates $(1, 2, 5)$ might be actively moved two units in the positive x direction, so that its new location becomes $(3, 2, 5)$. Alternatively, a passive transformation might be made, leaving the particle at $(1, 2, 5)$ right where it is, while shifting the coordinate system two units in the negative x direction, giving the untransformed location of the particle the new name $(3, 2, 5)$. In both cases, the state designated $(1, 2, 5)$ is transformed to the state named $(3, 2, 5)$. From the active point of view, the state is physically transformed, and the change of name simply reflects the fact. From the passive point of view, nothing is physically changed, and the change of name merely reflects the transformation of the coordinate system, the same state having different descriptions with respect to different coordinate systems. Note that the transformation of reference frame in the passive interpretation, a two-unit shift in the negative x direction, is the inverse of the transformation of state in the active interpretation, which is a two-unit shift in the positive x direction.

For another example, let us actively perform particle-antiparticle conjugation on a proton and change it into an antiproton, physically transforming the state called 'proton' to the state called 'antiproton', which is a very different state from 'proton'. (For instance, an antiproton carries negative electric charge, while a proton is positively charged.) Now instead, let us conjugate our bodies, our standard for distinguishing between particles and antiparticles, and transform our bodies into antibodies, while continuing to declare whatever is serving as our bodies to be the standard for the proton. Then the particle in state 'proton', without undergoing any physical transformation whatsoever, without changing its physical properties in the slightest, will find itself in state 'antiproton', since it will now be the antiparticle of the newly declared standard proton. It will have its untransformed state relabeled simply as a result of the passive transformation of reference frame.

Under both transformations a state designated 'proton' is transformed to a state called 'antiproton'. From the active point of view a proton is physically transformed into an antiproton, and the name of the state is changed accordingly. From the passive point of view there is no physical transformation, and the change of state designation is merely the result of the transformation of reference frame, changing the declared standard proton from a proton to an antiproton, with the same physical state having different names with respect to the different reference frames. In the present example both the active transformation and the passive transformation are particle-antiparticle conjugation, and this transformation is inverse to itself.

One might, and with good reason, ask how a transformation of reference frame is made. Is another reference frame needed to define the transformation? The answer is that the original reference frame serves as reference for its own transformation. The transformation is first defined with respect to that reference frame and then the transformation is performed on it.

The transformations we have been considering in the preceding sections were all active transformations. However, since it is formally possible to give any active transformation a passive interpretation as the inverse transformation performed on the reference frame, all the symmetries we have been discussing can be formulated equivalently in terms of immunity to possible (passive) transformation of reference frame rather than in terms of immunity to possible (active) transformation of experiment, result, state, etc. Still, I prefer the active point of view for formulating the symmetries of nature, one reason being that, as in the last example, transformed reference frames are not always physically realizable.

Let us taste the flavor of the passive point of view of symmetry in practice. Recall from Sect. 3.1 that symmetry of evolution of quasi-isolated systems means that there is some transformation that, if applied to any possible evolution of the system, would result in another possible evolution, and if hypothetically applied to any impossible evolution, would result in another impossible evolution. For quantum phenomena it would result in an evolution having the same probability. That is the active point of view. The passive formulation of symmetry of evolution of quasi-isolated systems is that there is some transformation such that, if applied to any reference frame, any pair of evolutions of identical description, as referred one to the transformed and the other to the original reference frame, are either both possible or both impossible or, for quantum phenomena, both have the same probability. This can be expressed in terms of experiments and their results. Consider some transformation applied to any reference frame. If for all pairs of experiments of identical description, as referred one to the transformed and the other to the original reference frame, their results, similarly and respectively referred, are also of identical description, we have symmetry of evolution. Or expressed very compactly, the passive formulation of symmetry of evolution is that the same physics is found with respect to all pairs of reference frames related by some transformation.

As an example, the active formulation of symmetry of evolution under spatial displacements, as presented in Sect. 3.1, is this: Any two physical systems that are simultaneously in identical states, except for one being here and the other being there, will evolve into final states that are also simultaneously identical, except for one being here and the other there, respectively, for all heres and theres. The passive formulation of that symmetry is: For any pair of spatially displaced, but otherwise identical, reference frames, any two physical systems that are simultaneously in states of identical description, as referred one to one reference frame and the other to the other, will evolve into final states that are also of identical description, similarly and respectively referred. Or, expressed compactly: The same physics is found with respect to all pairs of reference frames that differ solely by spatial displacement.

For another example, from Sect. 3.1 we have this for the active formulation of symmetry of evolution under boosts (velocity changes): Any two physical systems that are simultaneously in identical states, except for one moving at constant velocity with respect to the other, will evolve into final states that are also simultaneously identical, except for one moving at the same constant velocity with respect to the other, respectively, for all velocities up to the speed of light. Now formulated passively: For any pair of otherwise identical reference frames of which one is moving at constant velocity with respect to the other up to the speed of light, any two physical systems that are simultaneously in states of identical description, as referred one to one reference frame and the other to the other, will evolve into final states that are also of identical description, similarly and respectively referred. Expressed compactly: The same physics is found with respect to all pairs of reference frames that differ solely by relative motion at constant velocity up to the speed of light.

From the passive formulations of these examples we derive another reason for preferring the active point of view over the passive for formulating the symmetries of nature: When expressed in full detail, passive formulations turn out to be somewhat more awkward than their respective active counterparts. On the other hand, passive formulations have very compact versions.

3.4 Global, Inertial, and Local Reference Frames

The reference frames we have been considering in this section and in preceding sections have tacitly been assumed to be *global reference frames*, in the sense that each investigation makes use of a single reference frame that is taken to be in force at all locations and for all time. For example, the same x, y, z coordinate axes would cover all space and be valid for all times. Or, the same standard proton would serve for distinguishing between proton and antiproton over all space and for all time. Active transformations would be performed with respect to such global reference frames. Passive transformations would be made on them globally, thus changing them into new global reference frames.

Let us now consider the kind of evolution known as *inertial evolution*. Throughout such evolution nothing physically interesting happens at all: Newton's first law rules (so objects move at constant velocities), objects preserve their identities, etc.; in short, no dynamics. Reference frames relative to which dynamics-free evolution does in fact appear inertial are called *inertial frames*. (There is circularity here, but we will not go into the issue.) For example, a free proton with constant spin direction and at rest is an inertial evolution. So is such a proton that is moving with constant velocity. Reference frames with respect to which those evolution descriptions are obtained for a free proton are inertial frames.

It follows from the definition of symmetry of evolution (see Sect. 3.1) that the transformations involved in it transform inertial evolution into inertial evolution, from the active point of view, or inertial frames into inertial frames, when considered passively. Thus, such transformations can justifiably be described as *inertial transformations*. (The same can be said for the transformations involved in partial, limited, or approximate symmetry, but only in those cases where such symmetry is valid.) For example, a boost (velocity change) transforms a free proton at rest to one moving with constant velocity, both being inertial evolutions, when viewed actively, or makes a proton at rest appear to be moving with constant velocity, when considered passively. So boosts are inertial transformations. Or, particle-antiparticle conjugation transforms a resting proton to a resting antiproton, again an inertial evolution, from the active point of view, while from the passive point of view a resting proton is made to appear to be a resting antiproton. So particle-antiparticle conjugation is an inertial transformation as well.

Symmetry of evolution, then, includes what can be called symmetry of inertia, or *inertial symmetry*. The latter is defined as the possibility of a transformation under which all inertial evolutions remain inertial (or only some subset of them, for inexact symmetry), from the active point of view, or under which all inertial frames (or again, some) remain inertial, from the passive point of view.

Now, the concept of global reference frame can be generalized to that of *local reference frame*, which is the assignment of an individual reference frame to every point of space-time. (The technical term for a network of local reference frames over space-time is 'frame bundle'.) For example, the local x, y, z coordinate axes could have different orientations at different locations and at different instants, or the declared standard proton might vary by particle-antiparticle conjugation from location to location and from instant to instant. The reference for such space-time variation of local reference frames is the local reference frames at neighboring points. Thus, the space-time dependence of local reference frames is conventionally taken to be smooth, and discrete variation, such as a violently varying standard proton, is excluded.

As an example, consider the physical quantity known as isospin. This is an abstract quantity that is useful for describing nucleons (i.e., protons and neutrons), pions, and other particles. A nucleon is represented in abstract two-dimensional isospin space, with one axis for a neutron state and one axis for a proton state. Intermediate states, called *superposition* states – neither pure proton states nor pure neutron states – are also possible, and they have various probabilities to be detected as a neutron or as a proton. In a global isospin reference frame a free nucleon, say a proton, would maintain its identity as it moved along or remained at rest. However, a network of local isospin reference frames might be such that a free nucleon, again say a proton, would appear to smoothly change its identity, converting into a neutron-proton superposition or into a pure neutron, as it moves along or at rest. Or for another example, with respect to a network of local x, y, z coordinate axes, a physically free particle might seem to violate Newton's first law by undergoing what appear to be strange changes in its direction of motion.

The special case of a network of local reference frames that do not vary in space and time essentially defines and is equivalent to a global reference frame in the sense we have been using. It might be necessary to add additional specifications, though, such as the location of the origin of global coordinate axes derived from a space-time-nonvarying network of local coordinate axes. As might be expected due to the essential self-reference involved, the treatment of local *spatial* reference frames and local *spatiotemporal* reference frames [of the kind used in relativity, such as Minkowski coordinates (Hermann Minkowski, German mathematician, 1864–1909)] is mathematically more complicated than that of local abstract reference frames such as isospin, but this point will not concern us here.

3.5 Gauge Transformation

The transformations that we have considered so far for global reference frames, whether involved in symmetry or not, can now reasonably be called *global transformations*. Of course, global transformations can be made on networks of local reference frames and will affect all the local reference frames in the same way. But it is useful to generalize the concept of global transformation to that of transformation affecting local frames differently from location to location and from instant to instant, which can be called *local transformation*. Thus a local transformation involves at least one space-time-dependent parameter, specifying the effect of the transformation on the local reference frame at every location and instant. The space-time dependence of the transformation parameters must be smooth to ensure smoothness of the transformed local reference frame network. A global transformation is obtained as a special case of a local transformation, when the transformation parameters are taken to be space-time-independent.

As an example, a local xy rotation might be performed on a network of local coordinate axes by rotating the axes in the local xy plane at each location through an angle that depends on the location and time. Or, a local isospin transformation might be made on a network of local nucleon isospin reference frames by rotating the local neutron-proton axes through a space-time-dependent angle.

At this point let us consider the term 'gauge' as it is commonly used in expressions such as 'gauge symmetry' and 'gauge theory'. 'Gauge' is synonymous with 'reference frame'. A tire pressure gauge, for example, is a reference frame for air pressure, allowing the difference between the air pressure in the tire and that of the atmosphere to be specified with respect to some standard, such as Pa (pascal), equivalent to N/m^2 (newton per square meter), kg/cm^2 (kilogram force per square centimeter), or psi (pound force per square inch). A global gauge is a global reference frame, such as identically calibrated tire pressure gauges being used at all locations and for all time. A local gauge is accordingly a local reference frame, like the use of pressure gauges whose calibration varies from location to location and at each location varies with time.

A passive transformation, a transformation of reference frame, can then be called a gauge recalibration. However, we will use the conventional term *gauge transformation*. Such a transformation might be global or local, but in common usage 'gauge transformation' implies the more general, local transformation, which could, of course, be global as a special case. In the tire pressure gauge context a (tire pressure) gauge transformation would in general be performed differently at different locations and instants, but could possibly be done in a global manner. Gauge transformations are certainly not inertial in general, as they typically transform inertial reference frames to noninertial networks of local reference frames.

Consider, for example, the application of an isospin gauge transformation to an inertial nucleon isospin reference frame with respect to which a free proton is moving at constant velocity. With respect to the resulting network of local isospin reference frames (different isospin reference frames at different locations, all changing in time), the proton is still moving at the same constant velocity, but appears to change its identity among various neutron-proton superpositions (intermediate states), possibly even posing as a neutron, as it moves along. That evolution does not appear inertial. Indeed, gauge transformations can be said to introduce fictitious interactions or forces, in the present case an 'interaction' causing what appears to be a changing nucleon state (see Fig. 3.1).

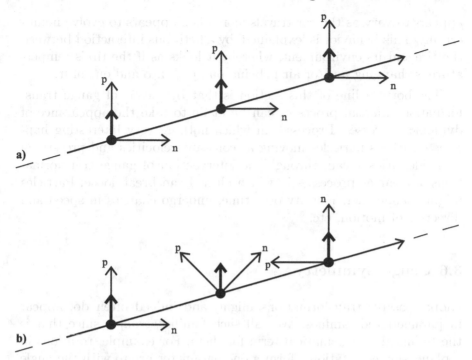

Fig. 3.1. (a) Free proton moving at constant velocity with respect to an inertial reference frame. (b) After an isospin gauge transformation, the same particle appears to change identity as it moves along

Or, let a gauge transformation involving space-time-dependent rotations be applied to an inertial reference frame with respect to which a free particle is moving with constant velocity, say a global set of x, y, z coordinate axes. Referred to the resulting network of local, time-dependent x, y, z coordinate axes, the particle appears to be in motion at constant speed and in continuously varying direction, which is accelerated, therefore noninertial, motion. Thus, this gauge transformation introduces fictitious forces 'responsible' for what appears to be accelerated motion.

As another example, let us start with a global tire pressure gauge, which is a set of identically and permanently calibrated tire pressure gauges distributed along the road. With respect to this global reference frame the pressure of a car tire does not vary as the car travels along. That is inertial evolution of tire pressure. Now let a gauge transformation be performed, recalibrating the pressure gauges differently at different locations along the road and in a time-dependent manner. With respect to the changed and changing gauges, the tire pressure appears to vary as the car travels along, i.e., appears to evolve noninertially. This behavior is 'explained' by a fictitious interaction between the tire and its environment, whereby it looks as if the tire's temperature is changing and/or air is being pumped into and out of it.

The bottom line of this section is that by means of gauge transformations inertial processes can be made to take the appearance of dynamic processes. Processes in which nothing very interesting happens – such as particles moving at constant velocities and preserving their identities – can, through the intervention of gauge transformations, appear as processes in which all hell can break loose: Particles might change their identity over time, undergo changes in speed and direction of motion, etc.

3.6 Gauge Symmetry

Global passive transformations might, and indeed often do, appear in parametrized families. We call such families *groups*, since that is the technical, mathematical term for them. For example, rotations in a plane, say xy rotations, form a one-parameter group with the angle of rotation serving as the parameter. Rotations of neutron-proton axes in abstract nucleon isospin space similarly constitute a one-parameter group. All rotations about a point make up a three-parameter group. Two of the parameters might specify the direction of the rotation axis, with the third determining the angle of rotation about this axis. Alternatively for those familiar with them, the Euler angles of rotation might serve as the three parameters.

Spatial displacements in all directions also form a three-parameter group of global passive transformations. In this case the three parameters could specify the values of the x, y, and z displacements needed to constitute the desired resultant displacement of the coordinate system. Boosts (velocity changes) in all directions form a three-parameter group as well. The values of the changes in the x, y, and z components of velocity that are required to achieve the boost might serve as the parameters.

Now, a parametrized group of global passive transformations can be converted into a group of (local) gauge transformations, can be 'gauged', as conventional jargon puts it. That is done by making the parameters of the group depend on location and time, by replacing the space-time-independent, global parameters with arbitrary space-time-dependent functions. Thus, the resulting local transformations, the resulting gauge transformations, will generally transform the local reference frames by different members of the original parametrized group at different locations and at different instants. Such a group is called a *gauge group*.

For example, under a member of the gauged group of rotations about a point acting on an ordinary global x, y, z coordinate system, at some instant and at some location the local coordinate system might be found rotated from its original orientation by $20°$ in the local xy plane, while at the same instant at some other location the local coordinate system might be found rotated by $12°$ in its yz plane. And at any fixed location the local coordinate system will generally vary its orientation with time, as the result of the time-dependent effect of the gauge transformation.

For another example, a gauge group can be obtained by gauging the group of nucleon isospin rotations. Then under a member of this group, at some location and at some instant the local neutron-proton axes might be found rotated from their original orientation by $22.5°$, while at the same location two seconds later they might be found rotated from their original orientation by $21.3°$. And at some other location the respective orientations at those instants might be $205°$ and $211°$ from the original orientation.

Let us now specifically consider parametrized groups of symmetry transformations of evolution (whether exact, partial, limited, or approximate), or in short, global symmetry groups, and let us zoom in on those groups of gauge transformations that are obtained by gauging global symmetry groups and which can thus be called global-symmetry-derived gauge groups. In spite of the inertiality of the transformations belonging to the original global symmetry groups (since, as symmetry transformations of evolution, they transform inertial evolution into inertial evolution), the gauge transformations thus obtained are not inertial, since by their local nature gauge transformations in general are not inertial, whatever the recipe used to cook them up. Thus these gauge transformations, just like all other gauge transformations, introduce fictitious forces and interactions in that they make inertial evolution appear noninertial.

We saw that in the three examples at the end of Sect. 3.5. The first example involves gauging the one-parameter group of rotations of the global nucleon isospin reference frame, which is related to a group of global partial symmetry transformations (that are valid only for certain classes of processes). The resulting gauge group introduces an 'interaction' causing what appear to be changes of neutron-proton identity. In the second example the three-parameter group of rotations of the global x, y, z coordinate axes, which is a global exact symmetry group, is gauged, introducing 'forces' responsible for what appears to be accelerated motion. And the third example is the gauging of the one-parameter group of global pressure scale changes, a global limited symmetry group, which introduces 'interactions' to explain what appears to be varying tire pressure.

Based on what we have seen so far, it would seem absurd to consider gauge groups as candidates for symmetries of evolution, even those obtained by gauging global symmetry groups – those we called global-symmetry-derived gauge groups – so that they have, one might say, symmetry in their genes. The inherent noninertiality of gauge transformations of any kind would seem to preclude this possibility. Nevertheless, it turns out to be very useful to consider global-symmetry-derived gauge groups as possible symmetries of evolution, and that happens because of, rather than in spite of, their noninertiality. The point is that nature seems to possess this symmetry, called *gauge symmetry*. More specifically, the evolution of quasi-isolated systems seems to be symmetric under certain global-symmetry-derived gauge groups. But not only that. When gauge symmetry is imposed on the theories that are proposed to explain the evolution of quasi-isolated systems, it constrains the theories so stringently that the dynamics, or at least its mathematical form, is essentially determined. The term 'gauge theory' refers to just such a gauge symmetric theory. So rather than being an absurd idea, gauge symmetry seems to be an extremely important idea.

What, then, is gauge symmetry? Since gauge transformations are passive transformations, we now take the passive point of view for symmetry of evolution of quasi-isolated systems. Gauge symmetry is the existence of a global-symmetry-derived gauge group – a group of gauge transformations obtained by gauging a group of global symmetry transformations – such that, if applied to any network of local reference frames, any pair of evolutions of identical description, as referred one to any transformed network and the other to the original network, are either both physically possible or both physically impossible or, for quantum phenomena, both possess the same probability. Or very compactly, we can say that gauge symmetry is that the same physics is found with respect to all local reference frame networks that are related by some global-symmetry-derived gauge group. That is the way gauge symmetry is usually defined, and that is the standard guidance to its application in physics (more specifically, in field theory).

But to obtain an idea of the *essence* of gauge symmetry rather than only its formal statement, we should turn from the passive to the active point of view. First we need the active interpretation of a gauge transformation: Consider an evolution described with respect to some network of local reference frames. Consider the action of a gauge transformation on the network. Referred to the transformed local reference frame network, the same evolution will have a transformed description. Reinterpret the transformed description as the description of a *transformed evolution with respect to the original network of local reference frames*, and this transformed evolution is the result of an active gauge transformation acting on the original evolution, the inverse of the passive gauge transformation that acted on the frame network.

For examples, we return again to the three at the end of Sect. 3.5. In the first example the original, inertial evolution is a free proton moving at constant velocity. A passive isospin gauge transformation is applied to the (inertial) global nucleon isospin reference frame. With respect to the resulting (noninertial) network of local nucleon isospin reference frames, the original evolution appears as a constant-velocity nucleon that is continuously changing its identity among neutron-proton superpositions. Thus, the effect of the active gauge transformation that is the inverse of the passive one is to transform the (inertial) constant-velocity proton evolution to the (noninertial) constant-velocity varying-identity nucleon evolution, where both evolutions are referred to the original (inertial) global nucleon isospin reference frame.

In the second example the original, inertial evolution is a free massive particle moving with constant velocity. A passive rotation gauge transformation is applied to the (inertial) global x, y, z coordinate axes. With respect to the resulting (noninertial) network of local x, y, z coordinate axes, the original evolution appears to be a massive particle in accelerated motion – at constant speed and in continuously changing direction. So the effect of the corresponding active gauge transformation is to transform the (inertial) free massive particle evolution to the (noninertial) accelerated massive particle evolution, with both evolutions referred to the original (inertial) global x, y, z axes.

In the third example the original, inertial evolution is a tire rolling along with constant pressure. A passive gauge (recalibration) transformation is applied to the (inertial) global tire pressure gauge (which is a set of identically and permanently calibrated tire pressure gauges distributed along the road). Referred to the resulting (noninertial) network of local recalibrated tire pressure gauges, the original evolution appears to be a tire rolling along with continuously varying pressure. Thus, the effect of the corresponding active gauge transformation is to transform the (inertial) constant-pressure rolling to the (noninertial) varying-pressure rolling, where both evolutions are referred to the original (inertial) global tire gauge.

We can now present an active formulation of gauge symmetry: If there is a global-symmetry-derived gauge group (a group of gauge transformations obtained by gauging a group of global symmetry transformations) such that all (or some subset of) evolutions related by it (as an active gauge transformation) are either all physically possible or all physically impossible or, for quantum phenomena, all possess the same probability, we have gauge symmetry (or partial, limited, or approximate gauge symmetry). This prepares us for working in the active picture and discovering what is actually implied by gauge symmetry.

As we saw, active gauge transformations transform inertial evolutions to noninertial, dynamic ones. In this sense active gauge transformations can be said to introduce dynamics. Gauge symmetry connects inertial and dynamic (noninertial) evolutions by ruling that the physical possibility, impossibility, or probability of some dynamic evolution is the same as that of some inertial evolution. Thus, gauge symmetry deserves to be called dynamic symmetry in the same way symmetry under global transformations is called inertial symmetry. In fact, global symmetry is a special case of gauge symmetry. When the space-time dependence of global-symmetry-derived gauge groups is made trivially constant, the global symmetry group that was gauged is regained, the dynamics accordingly disappears and turns into inertial evolution, and the dynamic symmetry becomes inertial.

The interdependence of inertial and dynamic evolutions required by gauge symmetry imposes severe constraints on the dynamics. To see that, let us return yet again to the three examples at the end of Sect. 3.5, as they were analyzed a few paragraphs above. The first example was simplified for the sake of presentation. The actual global symmetry of the strong interaction is the eight-parameter color transformation group – denoted SU(3) in group theoretical language – acting on the color states of the quarks composing the strongly interacting particles such as nucleons and pions. In the context of the strong interaction a constant-velocity proton is indeed a possible inertial evolution. Thus any (noninertial) varying-identity evolution obtained from it by the action of a color gauge transformation must also be possible. So the identity changing interaction that was considered to be fictitious from the passive point of view must now be taken seriously; an interaction really must exist.

Hence we must recognize the incompleteness of our original picture consisting solely of a nucleon, and a fortiori solely of a proton, and admit the essential ingredient of additional entities interacting with the quarks comprising the nucleon, which are the gluons. In this way we obtain the picture of quarks emitting and absorbing gluons and quarks interacting with quarks by means of gluon exchange. In quantum chromodynamics (QCD), which is the field theory that is supposed to explain such goings-on, the imposition of color gauge symmetry, together with the various global symmetries, forces the inclusion of a gluon field along with the quark field and determines the mathematical form of the interaction.

The second example, too, was simplified. A fuller global symmetry group is the six-parameter Lorentz group (Hendrik Antoon Lorentz, Dutch physicist, 1853–1928) consisting of all rotations about the origin and all boosts up to the speed of light. Since constant-velocity motion of a massive particle is a possible inertial evolution, so should the motion obtained from it by any Lorentz gauge transformation be a possible evolution. Such motions are generally accelerated motions, and the accelerations should be attributed to real forces. Therefore, the picture of a free massive particle is incomplete, and an accompanying force field, the gravitational field, must be appended. The corresponding gauge theory is the general theory of relativity, viewed as a field theory in flat space-time. The more common view of general relativity as a theory of inertial motion in curved space-time, rather than noninertial motion in flat space-time, is completely equivalent. The equivalence has to do with the arbitrariness inherent in the splitting of physical reality into background geometry (inertia) and dynamics, but we will not go into that issue. Here too, as in the case of strong interaction, the general theory of relativity is essentially determined by the gauge symmetry. It is the simplest nontrivial theory that possesses gauge symmetry under Lorentz gauge transformations.

In similar vein, the other two fundamental interactions of nature, the electromagnetic interaction and the weak interaction – or in their unified form, the electroweak interaction – are also successfully described by gauge theories based on gauge symmetries. (These gauge symmetry groups are denoted U(1) for the electromagnetic interaction and SU(2) for the weak interaction.) The laws of these interactions are essentially determined by the gauge symmetries.

Since the one-parameter group of global pressure gauge recalibrations in the third example is an extremely limited symmetry, so should be the corresponding local gauge symmetry, symmetry under local gauge recalibrations. But as this is a whimsical example anyway, we will push on and find out the imagined implications of such a gauge symmetry. Now, tire rolling at constant pressure is a possible inertial evolution. So all varying-pressure motions obtained from it by active local gauge recalibration should also be possible evolutions, and the varying pressure should be attributed to real effects: heating and cooling by the environment and pumping air into and out of the tire. Therefore, the picture of a free tire is incomplete, and accompanying heat and air pumping mechanisms must be appended. The thermal aspect is well covered by radiation, conduction, and convection. The insertion of air into and its removal from the tire are easily accomplished by a reversible air pump connected to the tire via a suitably swiveled hose.

Gauge symmetries' essentially unique determination of the theories of nature's fundamental interactions leads one to wonder whether symmetry might underlie all of nature, whether symmetry might even form the foundational principle of the Universe. Now, *that* is quite a conceptual leap. Yet, it has the right feel to it, at least to me it does. And I am not alone in this. Greene, for example, states in Chap. 8 of [20]: "From our modern perspective, symmetries are the foundation from which laws spring." See also [21].

And as for gauge symmetry itself, what is it telling us about nature's indifference, about what is irrelevant to nature, over and above the indifference implied by global symmetry? When we take the step from symmetry under space-time-independent transformations to symmetry under space-time-dependent ones, what are we adding to our understanding of nature? I do not have an answer. I do suspect that we are being told something, not merely about matter and its interactions, but about space-time itself at a deep level of reality. However, it is not at all clear to me what that might be. Something about independence, or perhaps interrelatedness, of events occurring at different locations or at different times? About locality and nonlocality? Perhaps about some aspect of nature that we are not yet even aware of? What a fascinating mystery!

3.7 Symmetry and Conservation

There is more to be said about symmetry of evolution, or symmetry of the laws of nature, and that concerns *conservation*. There are a number of conservations, also called *conservation laws*, that hold for quasi-isolated systems. The most commonly known of them are conservation of energy, conservation of linear momentum, conservation of angular momentum, and conservation of electric charge. What is meant is that, if the initial state of any quasi-isolated physical system is characterized by possessing definite values for one or more of those quantities, then any state that evolves from the initial state will have the same values for the quantities.

For example, in a particle scattering experiment, where an accelerated particle (perhaps a proton) collides with another (a gold nucleus, say), the vector sum of the linear momenta of all the participating particles before the scattering has taken place equals the vector sum of the linear momenta of all the particles that emerge from the collision (such as protons, neutrons, pions, and light nuclei, from high-energy proton–gold scattering). In addition, the total electric charge of the particles before collision equals their total charge afterward. And similarly for total angular momentum and total energy. These conservations hold even when particles are produced or annihilated during the process!

It turns out that each conservation is intimately and fundamentally related to a global symmetry group of evolution, to a certain symmetry of the laws of nature. Although there is considerable theoretical understanding of that relation, there is still much room for further investigation of it. A theoretical discussion of the relation is beyond the scope of this book.

Note that *conservation is itself symmetry*. Since the value of a conserved quantity does not change with time, we have symmetry of evolution of quasi-isolated systems under temporal displacements with respect to the value of the conserved quantity. Moreover, *symmetry of the laws of nature under global transformation groups is itself conservation*. Not conservation of a physical quantity, but conservation nevertheless. For such symmetry to be related to conservation in the manner we have discussed in this section, it must be valid for all times. Thus, it is the validity of the symmetry that is conserved.

Let us now mention the symmetries of the laws of nature that are related to the conservations mentioned above and show, for three of the four, how each conservation can be derived analytically from its related symmetry for a simple mechanical system.

Conservation of electric charge is related to symmetry of the laws of nature under a group of certain global phase transformations [denoted U(1)], to the fact that the laws of nature are the same whatever the phase of the system. (Recall that this is an abstract quantum phase, not phase in the sense of 'liquid phase'.) The description of this symmetry and the derivation of charge conservation from it are considerably more complicated than for the other symmetries and their related conservations, so we forgo its description and the derivation of the conservation from it.

3.7.1 Conservation of Energy

Conservation of energy is related to symmetry of the laws of nature under the group of global temporal-displacement transformations, to the fact that the laws of nature do not change with time. That is called temporal homogeneity of the laws of nature. It means that for every physically allowed process, all processes that are identical to that one except for their occurring at different times are also allowed by nature.

Consider the nonrelativistic system of a single point particle of mass m moving in one dimension with a potential $U(x)$ that is a function only of the particle's coordinate x. The system's evolution is determined by Newton's second law of motion,

$$F = m\ddot{x} ,$$

where F denotes the force on the particle and a double dot denotes the second time derivative, so $\ddot{x}$ is the particle's acceleration. The force is obtained from the potential by

$$F = -\frac{dU(x)}{dx} .$$

The particle's evolution is determined by the resulting equation for its acceleration,

$$\ddot{x} = -\frac{1}{m}\frac{dU(x)}{dx} .$$

The mass is constant, the potential energy does not depend explicitly on time, and the second derivative is immune to temporal displacements, $t \to t +$ constant. So this equation retains its form under temporal displacements. Thus, the evolution of the system is temporal-displacement symmetric.

The total energy of the system, the sum of kinetic and potential energies, is

$$E = \frac{1}{2}m\dot{x}^2 + U(x) \,,$$

where a single dot represents the first time derivative, so $\dot{x}$ is the particle's velocity. The time rate of change of the total energy is given by

$$\frac{dE}{dt} = m\dot{x}\ddot{x} + \frac{dU(x)}{dx}\dot{x} \,.$$

Substitute the acceleration $\ddot{x}$ from the previous paragraph in this expression to obtain

$$\frac{dE}{dt} = -\dot{x}\frac{dU(x)}{dx} + \frac{dU(x)}{dx}\dot{x} = 0 \,.$$

So the time rate of change of the total energy of this system vanishes, meaning that the total energy does not change with time; it is conserved.

3.7.2 Conservation of Linear Momentum

Conservation of linear momentum is related to symmetry of the laws of nature under the group of global spatial-displacement transformations, to the fact that the laws of nature are the same everywhere. That is referred to as spatial homogeneity of the laws of nature and means that for every physically allowed process all processes that are identical to it but occur at different locations are also allowed by nature.

As an example, consider a nonrelativistic one-dimensional system comprising of a number of point particles, labeled by i, j, k, whose masses are m_i. Let x_i denote the coordinate of the ith particle, $\dot{x}$, its velocity, and $\ddot{x}$, its acceleration. And let the particles interact via a potential U that depends only on the differences of the particles' coordinates. We now find the equation for the acceleration of the ith particle, the equation that governs the motion of that particle. Newton's second law of motion for the ith particle is

$$F_i = m_i\ddot{x}_i \,,$$

where F_i denotes the force on the particle. The force is obtained from the potential by

$$F_i = -\frac{\partial U}{\partial x_i} = -\sum_{j \neq i} \frac{\partial U}{\partial (x_i - x_j)} \frac{\partial (x_i - x_j)}{\partial x_i} = -\sum_{j \neq i} \frac{\partial U}{\partial (x_i - x_j)} \ .$$

The equation for the i th particle's acceleration is then

$$\ddot{x} = -\frac{1}{m_i} \sum_{j \neq i} \frac{\partial U}{\partial (x_i - x_j)} \ .$$

Neither the potential, its derivatives, the mass, nor the acceleration depends on the location of the system relative to the coordinate system's origin. So this equation is invariant under spatial displacements, $x_k \rightarrow x_k +$ constant. Thus, the evolution of the system is spatial-displacement symmetric.

The total linear momentum of the system is

$$P = \sum_i m_i \dot{x}_i \ .$$

The time rate of change of the total momentum is

$$\frac{\mathrm{d}P}{\mathrm{d}t} = \sum_i m_i \ddot{x} \ .$$

Now substitute the acceleration $\ddot{x}$ from the previous paragraph to obtain

$$\frac{\mathrm{d}P}{\mathrm{d}t} = -\sum_i \sum_{j \neq i} \frac{\partial U}{\partial (x_i - x_j)} = 0 \ .$$

The expression vanishes, since each pair of terms in the double sum with i and j interchanged cancel. Thus, the total linear momentum does not change in time; it is conserved.

3.7.3 Conservation of Angular Momentum

Conservation of angular momentum is related to symmetry of the laws of nature under the group of global rotations about all axes through a point, to the fact that the laws of nature are the same in all directions. That is called isotropy of the laws of nature. It means that for every physically allowed process, all processes that are identical to that one but have different spatial orientations are also allowed by nature.

Consider the nonrelativistic system of a single point particle of mass m moving in a plane in a central potential, i.e., a potential that depends only on the particle's distance from the coordinate origin, $U(r)$. Let (x, y) denote the particle's coordinates in the plane. Then its distance from the origin is

$$r = \sqrt{x^2 + y^2} \ .$$

The x-component of the particle's velocity is $\dot{x}$, the y-component of its velocity is $\dot{y}$, and the x- and y-components of its acceleration are $\ddot{x}$ and $\ddot{y}$, respectively. Newton's second law of motion for the particle takes the form

$$F_x = m\ddot{x} \ , \qquad F_y = m\ddot{y} \ ,$$

where F_x and F_y denote the x- and y-components of the force acting on the particle. The force components are obtained from the potential:

$$F_x = -\frac{\partial U(r)}{\partial x} = -\frac{dU(r)}{dr}\frac{\partial r}{\partial x} = -\frac{x}{r}\frac{dU(r)}{dr} \ ,$$

$$F_y = -\frac{\partial U(r)}{\partial y} = -\frac{dU(r)}{dr}\frac{\partial r}{\partial y} = -\frac{y}{r}\frac{dU(r)}{dr} \ .$$

The equations for the components of the particle's acceleration, which govern the particle's motion, are then

$$\ddot{x} = -\frac{x}{mr}\frac{dU(r)}{dr} \ ,$$

$$\ddot{y} = -\frac{y}{mr}\frac{dU(r)}{dr} \ .$$

Since the two equations, in x and in y, have the same form and since r is not affected by rotations about the origin, then any such rotation will result in a set of equations having the same form as this set, and the evolution of the system is rotation symmetric.

The particle's angular momentum with respect to the origin is

$$M = m(x\dot{y} - y\dot{x}) \ ,$$

and the time rate of change of the angular momentum is

$$\frac{dM}{dt} = m(x\ddot{y} - y\ddot{x}) .$$

Now substitute the acceleration components from the previous paragraph to obtain

$$\frac{dM}{dt} = -\frac{1}{r}(xy - yx)\frac{dU(r)}{dr} = 0 .$$

Thus, the angular momentum does not change with time; it is conserved.

3.8 Symmetry at the Foundation of Physics

In Sect. 2.6 we pointed out the three essential manifestations of symmetry at the foundation of science:

1. reproducibility,
2. predictability,
3. reduction,

with analogy and objectivity, also symmetries, giving a helping hand. Since physics is a science, in fact the most fundamental of sciences, these same symmetry manifestations appear also at the foundation of physics. In the present chapter we saw that the way physics analyzes and attempts to comprehend nature involves three more realizations of symmetry, which are

4. symmetry of evolution,
5. symmetry of states,
6. gauge symmetry.

To these we add

7. symmetry inherent to quantum theory,

since quantum theory introduces additional symmetry that is inherent to the foundation of quantum theory and thus also to the foundation of physics. That is the subject of Sect. 3.9.

These seven manifestations of symmetry lie at the foundation of physics. Since physics underlies the other sciences, all this symmetry lies also at the foundation of science. By the end of Chap. 2, symmetry manifestations 1–3 were sufficient by themselves to convince us – I hope – that science is founded on symmetry, indeed that science *is* symmetry. And now we have *seven* symmetry manifestations! That should nail it down extra solidly as far as science is concerned. But what about nature? Not only do we *view* nature through symmetry spectacles, as discussed in Sect. 2.6, but we *understand* nature in the language of symmetry. Can we then justifiably claim that nature, too, is symmetry? I do not think we can, at least not yet. I do expect, however, that if and when deeper levels of reality are ever investigated, they will be found to involve symmetry in a very major way. Perhaps symmetry will turn out to be what those fundamental levels are, in fact, all about. In the final analysis, then – if indeed there will eventually be a final analysis – will symmetry be revealed as the foundational principle of the Universe? See, for instance, [20] and [21].

3.9 Symmetry at the Foundation of Quantum Theory

We will see that much symmetry lies at the foundation of quantum theory. That symmetry also lies at the foundation of physics, since quantum theory forms a very fundamental part of physics. However, due to the very technical nature of a discussion of the symmetry, I prefer to relegate its discussion to the end of the chapter. The discussion assumes that the reader is sufficiently familiar with the Hilbert space formalism of quantum theory. The section may be skipped by those who are not sufficiently familiar.

3.9.1 Association of a Hilbert Space with a Physical System

This postulate has various formulations. It says that an abstract Hilbert space (David Hilbert, German mathematician, 1862–1943) is associated with every physical system. Furthermore, it says that to each state of a physical system there corresponds (or belongs, or that each state of a physical system is characterized by) a ray (all multiples of a vector), or a unit ray (all normalized multiples of a vector), in the Hilbert space associated with the system. That correspondence (or belonging, or characterization) is postulated to be exhaustive in the sense that there is no aspect of the state that is not comprehended by the ray, or unit ray. Moreover, every vector, or unit vector, of the

Hilbert space is postulated to correspond to (or belong to, or characterize) a state of the system. (The existence of superselection rules moderates this statement, but we will not get that technical.) Much symmetry is hiding within this postulate:

1. Since it is the *abstract* Hilbert space that is associated with a physical system, the physical significance of the association is immune to possible change of representation of the Hilbert space. For example, one might use coordinate representation, momentum representation, Heisenberg representation (Werner Karl Heisenberg, German physicist, 1901–1976), Schrödinger representation (Erwin Schrödinger, Austrian physicist, 1887–1961), etc. The same physics is being described in each case.

2. If it is a ray in the Hilbert space that corresponds to a state of the system, then given a vector corresponding to a state, all multiples of the vector correspond to the same state. Thus the vector → state correspondence is immune to possible multiplication of the vector by any complex number. If unit ray, rather than ray, serves in the formulation, then the correspondence is immune to possible multiplication of the unit vector by any phase factor, i.e., a complex number of unit modulus. The moduli and phases of all state vectors anyway cancel out in the formulas for physical results (i.e., probabilities and expectation values, see Sect. 3.9.7 below). Thus the modulus and phase of a vector in a system's Hilbert space are irrelevant to quantum theory. That does not exclude their potential relevance to a more general theory than quantum theory.

3. The exhaustiveness of the vector → state correspondence does not mean that a system's states cannot possess aspects that are not comprehended by vectors of the system's Hilbert space. It does mean, however, that if a system's states do possess such aspects, those aspects are irrelevant to quantum theory. Thus the physical properties and evolution that are determined by quantum theory for a state are immune to possible change in such aspects of the state. Obvious examples of such aspects are the esthetic or moral value one might assign to a state. And there might even be objective aspects, such as of the nonlocal-hidden-variable variety, that, in spite of their irrelevance to quantum theory, might be relevant to a more general theory.

3.9.2 Correspondence of Observables to Hermitian Operators

Another postulate of quantum theory states that all observables (physical quantities) of a system correspond to Hermitian (self-adjoint) operators (Charles Hermite, French mathematician, 1822–1901) in the system's Hilbert space and that the possible values of an observable are the eigenvalues of its corresponding operator. As far as is known to me, this correspondence is unique for each observer; i.e., possibly barring some pathological cases that I have not heard about, for each reference frame there is but a single Hermitian operator corresponding to each observable. As an example of different observers having different operators for the 'same' observable, the addition of a constant real multiple of the identity operator to the energy operator of one observer gives the operator corresponding to the energy of the system relative to the shifted energy reference level of another observer.

1. This does not imply that the operator corresponding to some observable for a given system always has the same representation. Just as the physical significance of the association of an abstract Hilbert space with a system is immune to possible change of the representation of that space (see item 2 in Sect. 3.9.1 above), so is the correspondence of an observable to a Hermitian operator for a given system immune to possible change of the representation of that operator. The correspondence is with the *abstract* operator.

2. The possible values of an observable are the possible results of measuring that quantity. But the postulate does not specify just how the measurement is to be carried out. If we have a classical criterion for determining whether different measurement procedures accord with the same observable, then the observable → operator correspondence is immune to possible change in the measurement process, as long as the same observable is being measured. If there is no such criterion, then we are dealing with a purely quantal physical quantity, having no classical analog. In this case the observable will more or less be defined by its operator (eigenvalue spectrum), and all measurement procedures corresponding to the operator will compose the range of possible change under which the correspondence is immune, in an essentially tautological manner.

3. The possible values of an observable for a given system are postulated to be the eigenvalues of its corresponding Hermitian operator in the Hilbert space of the system. This postulate is immune to possible change in the result of measuring the observable, where the

range of possible change is the eigenvalue spectrum of the corresponding operator. In other words, the same measurement procedure might give different results, but all possible results are constrained to belong to the eigenvalue spectrum of the corresponding operator. This is much the same as in classical physics.

4. But in quantum theory the statement has an added twist: Even when the same measurement procedure is applied to the *same state* of the system, different results might be obtained, but all possible results are constrained to belong to the eigenvalue spectrum of the corresponding operator. (In classical physics, of course, all such results would simply be the same.)

3.9.3 Complete Set of Compatible Observables

A complete set of compatible observables (or variables) for a given system is a maximal set of simultaneously measurable functionally independent observables for the system. Such a set corresponds to a complete set of commuting Hermitian operators in the system's Hilbert space. A complete set of compatible observables can in general be chosen in various ways for a given system, and each choice corresponds to a different complete set of commuting Hermitian operators.

1. Each choice of a complete set of compatible observables implies a different way of specifying a state of the system. The physical properties and evolution determined by quantum theory for a state are immune to possible change in the way the state is specified, and thus also to possible change in the choice of complete set of compatible observables.

2. Modulo degeneracy complications, the simultaneous eigenvectors of a complete set of commuting Hermitian operators can serve as a basis for the Hilbert space, giving a representation of the space along with its vectors and operators. The physical significance of the association of a Hilbert space with a system, along with the correspondence of vectors, or unit vectors, to states and of Hermitian operators to observables, is thus immune to possible change in the choice of complete set of compatible observables (see item 1 in Sect. 3.9.1 and item 1 in Sect. 3.9.2 above).

3.9.4 Heisenberg Commutation Relations

To quantize a classical system described in terms of a set of canonical variables, the Hermitian operators corresponding to the canonical variables are postulated to obey the Heisenberg commutation relations.

The same classical system can be equivalently described by different sets of canonical variables related to each other by canonical transformations. In quantizing the system, the same Heisenberg commutation relations are imposed on the Hermitian operators corresponding to whatever set of canonical variables one uses. Thus, the Heisenberg commutation relations for the Hermitian operators corresponding to the canonical variables are immune to possible canonical transformation of those variables. This symmetry follows from the invariance of the Poisson bracket relations (Siméon Denis Poisson, French mathematician and physicist, 1781–1840) for the canonical variables themselves under canonical transformations, which we can express analogously as the immunity of the Poisson bracket relations for the canonical variables to possible canonical transformation of those variables.

3.9.5 Operators for Canonical Variables

For a classical system described in terms of a set of canonical variables, physical quantities are expressed as functions of the canonical variables. Modulo ordering technicalities (such as symmetrization, time ordering, normal ordering), the Hermitian operators corresponding to observables for the quantum analog of a classical system are obtained from the above-mentioned functions by substituting for the canonical variables their corresponding operators.

Different sets of canonical variables, related to each other by canonical transformations, can be used to equivalently describe a classical system. Different sets of canonical variables entail different functions for the same physical quantity and hence different forms for the Hermitian operator corresponding to the same observable. Thus, the physical significance of a Hermitian operator is immune to possible change in its form entailed by canonical transformation of the canonical variables.

3.9.6 A Measurement Result Is an Eigenvalue

Any vector in the Hilbert space of a system can be decomposed into a linear combination of eigenvectors of any Hermitian operator in the space. Accordingly, any state of the system can be considered as a superposition of eigenstates of any observable (with appropriate qualification in the case of superselection rules, as pointed out in Sect. 3.9.1 above). It is postulated that a measurement of an observable indeterministically brings about a change of the system's state to one of the observable's eigenstates, of which the original state can be considered a superposition, i.e., the system's state is indeterministically projected onto one of the set of eigenstates of the measured observable. The result of the measurement is postulated to be the eigenvalue corresponding to the resulting eigenstate.

The symmetry here is that the general effect of a measurement, as describe above, is independent of the state in which the system actually is. *Whatever* the state of a system, the measurement of an observable indeterministically projects that state as described above, bringing about some eigenstate of the measured observable and giving the corresponding eigenvalue as the measurement result. Thus, the effect of a measurement (but not its numerical result!) is immune to possible change of the state of the system.

3.9.7 Expectation Values and Probabilities

The formulas for expectation values and probabilities are invariant under multiplication of each vector appearing in them by any complex number (which may be different for each vector), if 'ray' is used in Sect. 3.9.1 above, or by any phase factor (which may be different for each vector), if 'unit ray' is used in the same section. So calculated expectation values and probabilities are immune to the choice of vector from the ray, or unit ray, corresponding to a state of the system (see item 1 in Sect. 3.9.2 above).

3.9.8 The Hamiltonian Operator

The Hamiltonian operator (William Rowan Hamilton, Irish mathematician and physicist, 1805–1865), the Hermitian operator corresponding to the energy of the system, is the generator of the system's deterministic evolution.

This symmetry is similar to that of Sect. 3.9.6. *Whatever* state the system may be in, the Hamiltonian is the generator of the deterministic temporal evolution of the system from that state to the future and from the past to that state. The Hamiltonian's property of being the generator of deterministic evolution is immune to possible change of the state of the evolving system.

3.9.9 Planck's Constant as a Parameter

In the mathematical formalism of quantum theory Planck's constant h (Max Karl Ernst Ludwig Planck, German physicist, 1858–1947) appears as a parameter. Only experiment determines the actual value of h in terms of conventional units (or alternatively, determines conventional units in term of h and the other fundamental constants).

Thus, the mathematical formalism of quantum theory possesses the symmetry that the formalism (but not values of calculated results!) is immune to possible change of the actual value of h.

3.9.10 The Correspondence Principle

For a quantized system with classical analog, the expectation values of Hermitian operators corresponding to observables must behave in the limit $h \to 0$ like the respectively corresponding classical physical quantities. (However, in certain cases the correspondence principle breaks down.)

1. This principle is the particular application to quantum theory of the general correspondence principle, that *any* physical theory more general than classical theory must be consistent with classical theory in the latter's domain of validity, that such a theory must in some sense reduce to classical theory in that domain. For the particular case of quantum theory, the reduction is carried out by taking the limit $h \to 0$ in the mathematical formalism of the theory. The general correspondence principle therefore possesses the symmetry that its validity is immune to possible change of the more general physical theory to which it is applied.

2. The particular correspondence principle for quantum theory shares the symmetry presented in the preceding paragraph. Its validity is immune to possible change of the actual value of h.

3.10 Summary

We discussed symmetry of evolution of quasi-isolated physical systems and symmetry of states of such systems in Sects. 3.1 and 3.2, respectively. The former symmetry, also called symmetry of the laws of nature, is a manifestation of nature's indifference to certain aspects of physical systems. When systems differ only in such aspects, nature treats the systems in essentially the same manner so that they evolve in essentially the same way. States of systems are symmetric when changes can be applied to them that affect only those of their aspects to which nature is indifferent.

In Sect. 3.3 we discussed reference frames, which are needed to endow transformations with physical significance and must themselves be asymmetric under the transformations. The notion of active and passive transformations was introduced, where the former affect physical systems (and not reference frames), while the latter act only on reference frames (and not on systems). That led to active and passive formulations of symmetry of evolution. Both formulations involve different evolving physical systems. In the active formulation the evolutions have different descriptions with respect to the same reference frame, while in the passive formulation they have the same description with respect to different reference frames.

In Sect. 3.4 we introduced the concepts of global reference frame, which is a single frame that is valid for all space and all time; inertial reference frame, which is a reference frame in which inertial motion appears as such; and local reference frame, which is the assignment of an individual reference frame to every space-time point, i.e., the assignment of an individual time varying reference frame to every point in space.

We discussed gauge issues in Sects. 3.5 and 3.6. A global transformation is one that has the same effect at all locations and instants. Gauge transformations are passive transformations that in general have different effects at different locations and instants (although, as a particular case, a gauge transformation might have the same effect at all locations and instants, making it a global transformation). A gauge group is a group of gauge transformations that are obtained by taking a parametrized group of global transformations and making its parameters space-time-dependent. That leads to the notion of gauge symmetry, which is symmetry of evolution under a gauge group. Gauge symmetry relates inertial and dynamic evolutions. Gauge theories are theories that possess gauge symmetry. Indeed, the most successful theories of the fundamental particles and their interactions are gauge theories. Further, the gauge symmetries essentially determine these theories.

The relation between symmetries and conservations, also known as conservation laws, was discussed in Sect. 3.7. Each conservation is fundamentally linked to a symmetry of evolution, or symmetry of the laws of nature.

The symmetry that lies at the foundation of physics was presented in Sects. 3.8 and 3.9. It comprises the symmetry lying at the foundation of science – reproducibility, predictability, and reduction, with the help of analogy (all discussed in Chap. 2) – and the realizations of symmetry that are particular to physics – symmetry of evolution, symmetry of states, gauge symmetry, and the symmetry inherent to quantum theory.

In this chapter we used loosely a number of technical terms that possess precise definitions. They include transformation, group, symmetry transformation, and symmetry group and will be treated rigorously in Chap. 10. My purpose in taking this approach is to try to make the concepts as accessible as possible with only the absolute minimum of mathematics. However, for the complete picture the relevant mathematical background is needed and is presented in Chaps. 8–10.

4

The Symmetry Principle

The fundamental principle in the application of symmetry considerations to problem solving in science and engineering and devising theories in physics is the symmetry principle, also known as Curie's principle (Pierre Curie, French physicist, 1859–1906). In this chapter we will see how the symmetry principle follows from one of the most basic notions in all of science, indeed, one of the most crucial underpinnings of our grasp of reality – the causal relation.

4.1 Causal Relation

The first step in our development of the symmetry principle is achieving understanding of the notion of causal relation, also called cause-effect relation. It might seem that in everyday affairs there should be no difficulty in understanding that relation; the cause brings about the effect, and the effect is a result of the cause. Obvious, right? But is the situation really so clear? For example, at the end of a concert you applaud, whereby your clapping makes a noise that expresses your appreciation. The clapping makes the noise, so the clapping is the cause and the noise is the effect, which is satisfactory. But why do you clap? You clap to make the noise; if clapping did not make that kind of noise, you would not clap. Thus, your clapping is performed because of the noise. And does this not mean that the noise is the cause and the clapping the effect? Or is it the desire to produce the noise that is the cause?

Possibly in everyday affairs we can make do with such foggy concepts, but in science our concepts must be much clearer (which would do no harm in everyday affairs too!). So we will attempt to clarify the concept of causal relation to a degree of hairsplitting that is sufficient for our purposes.

Now, we have been using the term 'system' in this book to indicate a physical system. Since the symmetry principle is applicable to more than just physical systems, we will make its derivation as general as possible. To this end we will use 'system' in a more general way, to indicate whatever it is that attracts our interest and attention. Any part of a system is termed a subsystem. A subsystem of a system is also a system in its own right. And just as a system can have states, so can a subsystem. The state of a system determines the state of each of its subsystems. However, the state of a subsystem does not in general determine the state of the whole system or of any other subsystem (although that might happen in certain cases).

Consider two subsystems, which we denote A and B, of any system. Consider all states of the whole system and imagine that we set up a triple-column list of pairs of states of subsystems A and B that are determined by each state of the whole system for all of its states. Of course, it is possible that the same state of a subsystem will appear more than once in the resulting list (see Table 4.1).

Let us now look for a correlation between states of subsystem A and states of subsystem B in the list by asking the following questions: (1) Does the same state of subsystem A always appear with the same state of subsystem B (and possibly different states of A appear with the same state of B)? (2) Does the same state of subsystem B always appear with the same state of subsystem A (and possibly different states of B appear with the same state of A)? If the answer to either or both questions is affirmative, we say that a *causal relation*, or a *cause-effect*

Table 4.1. States of two subsystems determined by the state of the whole system

States of whole system	Determine states of	
	Subsystem A	Subsystem B
a	h	p
b	i	p
c	h	q
d	j	r
e	j	s
⋮	⋮	⋮

relation, exists between the two subsystems. If the first question is answered affirmatively, we say that subsystem A is a *cause subsystem* and subsystem B is an *effect subsystem*. If the second question is answered affirmatively, we say that B is a cause subsystem and A is an effect subsystem. If the answers to both questions are affirmative, each subsystem is both cause and effect (see Tables 4.2 and 4.3).

In that manner the definition of a causal relation between subsystems is based on a correlation between their states. There is not even a hint of 'bringing about', 'resulting', 'producing', or 'causing' (in the usual sense). The states of one subsystem and those of the other are connected through the fact that they are determined by the states of the same whole system, and this connection is the origin of possible causal relations between subsystems. If we think we understand the 'mechanism' that underlies the correlation, we express the relation in

Table 4.2. A is a cause subsystem and B is an effect subsystem

States of whole system	Determine states of	
	Subsystem A	Subsystem B
a	h	p
b	i	q
c	j	r
d	i	q
e	l	r
f	m	s
⋮	⋮	⋮

Table 4.3. Each subsystem is both a cause and an effect subsystem

States of whole system	Determine states of	
	Subsystem A	Subsystem B
a	h	p
b	i	q
c	j	r
d	k	s
e	h	p
f	l	t
⋮	⋮	⋮

terms of 'producing', etc. But even if we do not understand the 'mechanism' or perceive any 'mechanism' at all, a causal relation may still
exist between systems as an empirical fact of correlation between their
states. Upon discovering a pair of causally related systems, we do indeed tend to search for a whole system of which the causally related
systems form subsystems, in order to understand the causal relation.
However, the existence of a causal relation and our knowledge of its
existence in no way depend on our understanding or perceiving its
underlying 'mechanism'.

And how does that abstract definition of causal relation relate to
our intuitive understanding of the concept of causal relation? First
of all, I would like to warn against attaching too much importance
to one's intuitions, at least in science. Intuitions are largely thought
habits. And since those habits developed as a result of limited experience, their appropriateness to phenomena lying outside this range of
experience is suspect, at the least. (Need I remind you of the history of
the special theory of relativity and quantum theory?) Second of all, in
spite of that – I merely wanted to express my opinion and offer a warning – there is no contradiction between our abstract definition of causal
relation and our intuitive understanding of the concept. We are used
to the idea that, if a cause produces an effect, every time the cause is
in the same state the effect is in the same state. Otherwise we would
not have even considered the 'cause' as being a cause to begin with.

For example, we are used to thinking that forces produce acceleration, that the forces are the cause and the acceleration is the effect,
and not vice versa. And indeed, whenever the same set of forces acts
on a given body, the same acceleration occurs. We cannot agree that
the acceleration is the cause and the forces the effect, because the same
acceleration may be correlated with different sets of forces. Different
sets of forces may be correlated with the same acceleration, while different accelerations are never correlated with the same set of forces.
In abstract terms the system comprises the body, the forces acting on
it, and its acceleration. The forces form subsystem A and the acceleration makes up subsystem B. The answer to the first question – Does
the same state of subsystem A always appear with the same state of
subsystem B (and possibly different states of A appear with the same
state of B)? – is affirmative; the same state of forces is always correlated with the same state of acceleration, and even different states of
forces are correlated with the same state of acceleration. The answer
to the second question – Does the same state of subsystem B always
appear with the same state of subsystem A (and possibly different

states of B appear with the same state of A)? – is negative; the same state of acceleration is not always correlated with the same state of forces. Therefore, the forces and the acceleration are in causal relation, with the forces the cause and the acceleration the effect.

Let us return to our example of applause. The system consists of the human body, all its actions, and all its noises. The hands compose subsystem A, which is considered to possess only two states: clapping and not clapping. Subsystem B is composed of all body noises and is also considered to have only two states: clapping noise and not clapping noise (i.e., silence or any other noise). As we review all states of the whole system, we find a most astounding correlation between the states of A and the states of B: Whenever the hands clap, clapping noise sounds; and whenever clapping noise sounds, the hands are clapping. Thus, the answers to both the first and the second questions are affirmative, and there exists a causal relation between hand clapping and clapping noise, where each may be taken as cause, as effect, or as both.

Another example is the space-time configuration of electric charges and currents as the cause and the electric and magnetic field strengths as effect. Or, for a quantum system we could have the scattering potential and incident wave function as cause and the scattered wave function as effect. The cause might be your present situation and recent history, with your present mood as effect. And so on.

Or consider the famous Rutherford experiment, in which alpha particles are scattered from stationary metal nuclei. If the exact locations of the target nuclei and the precise trajectory of each incident particle could be known, those would serve as the cause, and the effect would be the point where each scattered alpha particle hits the fluorescent detection screen. However, neither the locations of the target nuclei nor the trajectory of each incident particle are known. The best we can do is to confine the metal nuclei to a thin foil of known density and to know the statistical properties of a beam of alpha particles (particle flux, velocity distribution, etc.). Taking these as the cause, the effect certainly cannot be the final trajectory of a scattered particle or even its scattering angle, since scattering in all directions is actually observed in the experiment. The effect is, in fact, the statistical distribution of scattering angles for the scattered alpha particles, expressed by the angular-distribution density function or the differential scattering cross-section.

4.2 Equivalence Relation, Equivalence Class

The ultimate goal of our line of reasoning that started with a discussion of causal relation is the symmetry principle. Along that line we must stop at three way stations, which cannot be avoided if we want not only to use the symmetry principle but to understand it. At our first stop along the way we must become acquainted with the concept of *equivalence relation* for states of a system. Actually, the concept of equivalent relation is more general than only for states of a system and is applicable to any set of elements, whatever their nature might be. So here we will introduce the concept for an abstract set of abstract elements, and later we will apply it to sets of states of systems. An equivalence relation for a set of elements is defined as any relation, commonly denoted by $\equiv$, that might hold for pairs of elements of the set and satisfies these three conditions:

1. *Reflexivity*. Every element of the set has this relation with itself, i.e.,

$$a \equiv a \,,$$

   for all elements a of the set.

2. *Symmetry*. If one element has the relation with another, then the second has it with the first, for all elements of the set. In symbols that is

$$a \equiv b \iff b \equiv a \,,$$

   for all elements a, b of the set. (The arrow $\Rightarrow$ denotes implication in the arrow's direction. The double arrow $\Leftrightarrow$ indicates implication both ways. Recall that whatever stands at the head of an implication arrow is a necessary condition for whatever stands at the tail. And whatever stands at the tail is a sufficient condition for whatever stands at the head. Whatever stands at one head of a double implication arrow is a necessary and sufficient condition for whatever stands at the other head.)

3. *Transitivity*. If one element possesses the relation with a second element and the second has it with a third, then the first element has the relation with the third, for all elements of the set. Represented symbolically, that is

$$a \equiv b \,, \quad b \equiv c \implies a \equiv c \,,$$

   for all elements a, b, c of the set.

One somewhat fanciful example of an equivalence relation is friendship. (1) It is a reflexive relation, since everyone is presumably a friend of himself or herself. (2) It is symmetric, since if you are my friend, I am your friend. (3) It is transitive, since a friend of a friend is (in an ideal world) a friend.

More seriously, the most familiar example of equivalence relation is the relation of equality, as denoted by =. (1) Every number is equal to itself. (2) If number a equals number b, then b equals a. (3) If a equals b and b equals c, then a equals c. Quite obvious.

Similarly, the congruence relation of modular arithmetic of integers, written $\equiv$ (mod n), is an equivalence relation. And so is the geometric relation of similarity (possessing the same shape) for figures in a plane. For another example of equivalence relation, take the relation among people of having the same birthday. And yet another, the relation among purebred dogs of being of the same breed.

An example of a relation that is not an equivalence relation is the relation of 'less than' between numbers, denoted $<$. (1) It is not reflexive, since no number is less than itself. (2) Neither is it symmetric, since if number a is less than number b, certainly b is not less than a. (3) The relation is, however, transitive; if a is less than b and b is less than c, then a is less than c.

Just for practice, let us modify the 'less than' relation to 'less than or equal to', which is denoted $\leq$. (1) This relation is reflexive, since every number fulfills it with itself thanks to the 'equal to' option. (2) But the addition of the 'equal to' option does not make the relation symmetric. If a is less than or equal to b, then it does not follow in general that b is less than or equal to a. That would happen only in the special case that $a = b$, but not in general. (3) Transitivity is fulfilled.

Another example is inequality, denoted $\neq$. (2) This relation is symmetric, but is neither (1) reflexive nor (3) transitive. (If a does not equal b and b does not equal c, nothing is implied about the relation between a and c.)

In a set of elements (which might be a set of states of a system) for which an equivalence relation is defined, a complete set of mutually equivalent elements is a subset of elements all of which are equivalent with each other and only with each other. Such a subset is called an *equivalence class*. No element of the set can simultaneously be a member of two different equivalence classes. If one hypothetically were, then due to the symmetry and transitivity properties of an equivalence relation, what we thought were two equivalence classes would really form a single equivalence class. Thus, an equivalence relation brings about a decomposition of the set of elements for which it is defined into equivalence classes. Every element of the set is a member of one and only one equivalence class. And any element that happens to be equivalent only with itself simply forms an equivalence class of one.

As an example, consider the equality relation among numbers. Every number is equal only to itself. So the equivalence classes contain only a single number in each.

For a more interesting example, take the congruence relation $\equiv$ (mod n) of modular arithmetic of integers. This equivalence relation is that two integer numbers are equivalent, or congruent, when they leave the same remainder when divided by the integer n. For instance, $1 \equiv 13$ (mod 4). This particular relation decomposes the set of all integers into four equivalence classes: (0) the class of all multiples of 4, $\{\ldots, -8, -4, 0, 4, 8, \ldots\}$, all integers that leave a remainder of 0 when divided by 4; (1) the class of all integers that leave a remainder of 1 when divided by 4, $\{\ldots, -7, -3, 1, 5, 9, \ldots\}$; (2) the class $\{\ldots, -6, -2, 2, 6, 10, \ldots\}$, all of which leave a remainder of 2; and (3) the class $\{\ldots, -5, -1, 3, 7, 11, \ldots\}$ of all integers leaving a remainder of 3. All numbers within each equivalence class are equivalent (congruent) with each other, and no number is equivalent (congruent) with a number in a different equivalent class.

As a further example, consider the geometric relation of similarity (possessing the same shape) for figures in a plane. This equivalence relation decomposes that set of all plane figures into equivalence classes, each of which contains all figures of the same shape and different size. For instance, one equivalence class contains all equilateral triangles, another contains all 30°-60°-90° triangles, another all squares, yet another all rectangles whose sides are in 1:3 proportion, and so on and so forth.

Conversely, any decomposition of a set of elements (possibly a set of states of a system) into subsets, such that every element of the set is contained in one and only one subset, defines an equivalence relation. The subsets may then be declared, by fiat, equivalence classes, and the corresponding equivalence relation is simply that two elements are equivalent if and only if they belong to the same subset.

The decomposition of humanity into 366 subsets of people having the same birthday serves as an example of that. This decomposition defines the equivalence relation of having the same birthday. Thus all people whose birthdays fall on 16 November, for instance, are thus defined as equivalent. And no such person is equivalent with anybody whose birthday is 9 July.

Similarly, the decomposition of the population of purebred dogs into pure breeds as subsets defines the equivalence relation of being of the same breed. All miniature poodles, for instance, are thus equivalent.

As another example, the decomposition of the human population into eight subsets characterized by blood class (A+, A−, B+, ..., O−) defines an equivalence relation for all humans, the relation of having blood of the same class.

That leads us back to Sect. 1.3, where it was concluded that analogy and classification are symmetry. Indeed, analogy and classification impose just such decompositions on sets and thus define equivalence relations. The blood class example was presented in that section as well. The other examples can similarly be immediately translated from the language of analogy and classification into the language of equivalence and equivalence class.

I might point out that every set of elements (which could be a set of states of a system) possesses two trivial decompositions into equivalence classes. The most exclusive equivalence relation, that every element is equivalent only with itself, decomposes the set into as many equivalence classes as there are elements of the set, since every equivalence class contains but a single element. We saw that in the example of the equivalence relation of equality for the set of all numbers.

On the other hand, the most inclusive equivalence relation, that all elements of the set are equivalent with each other, makes the whole set a single equivalence class in itself. For instance, we can enhance our feeling of unity with all of humankind by recognizing that the equivalence relation among people that all human beings are equivalent determines that the whole human population forms a single equivalence class.

4.3 The Equivalence Principle

We now have two more intermediate stops along the way to the symmetry principle. At the first of them we must clarify to ourselves certain points concerning the nature of science. At the second stop we will show that the very existence of science implies the *equivalence principle*, which is: *Equivalent causes – equivalent effects*. We will deal with those two issues in this section. In Sect. 4.4 we will derive the long-awaited symmetry principle as an almost immediate consequence of the equivalence principle, and we defer its formulation to there.

One component of the foundation of science is reproducibility (about which, see Sects. 2.1 and 2.3). Reproducibility is the possibility of repeating experiments under conditions that are similar in certain respects, yet different in other respects (such as location or time), and obtaining essentially the same results. (Section 2.3 goes into more detail.) That possibility furnishes the objectivity that is essential for science to be a common, lasting endeavor rather than a set of private sciences or impossible altogether. We are not claiming that all the phenomena in the world fulfill the condition of reproducibility. For instance, the phenomena of parapsychology are notorious for their irreproducibility. This fact does not negate the possible existence of such phenomena nor does it necessarily invalidate the investigation of them. But as long as reproducibility is not achieved in parapsychology, the latter will continue to lie outside the domain of concern of science.

Another component of the foundation of science is predictability (discussed in Sects. 2.1 and 2.4). Predictability is the possibility of predicting the result of as yet unperformed experiments. This means that, until proved otherwise, we labor under the assumption that human intelligence is capable of understanding nature sufficiently to allow the prediction of phenomena that have not yet been observed. Reproducibility alone, without predictability, does not make science. It only allows recording, cataloging, and classifying of experimental observations as public information, with no benefit beyond the compilation itself. Science begins to be possible only when order is perceived among the collected facts, and on the basis of that order the results of new experiments are predicted. (See Sect. 2.4 for more detail.)

The scientific tool by which we predict the results of experiments is the law. A law is any conceptual recipe or mathematical formula or other means like those that, when fed data about the conditions of an experiment, tells us the experimental result to be expected. The first test of a proposed law is performed by comparing it with past experiments and their results. If the results expected according to the law match the results actually obtained, the law advances to the next testing stage. If not, it goes back to the drawing board for corrections or overhaul or into the wastebasket. For the next test the law must predict the results of as yet unperformed experiments. If it does that with continuing success, our confidence in it continually increases. However, a single failure is sufficient to invalidate a proposed law in spite of its past success.

Reproducibility and predictability are expressions of our realization that causal relations exist in nature. A scientific law is the expression of a particular causal relation: The data that the law receives represent the physical cause, and the results that the law gives represent the physical effect. When a law receives the same data, it always gives the same results, and that reflects the relation between the cause and the effect.

A scientific law may be considered as including a sort of antianalog computer. I use the term 'antianalog computer', since, whereas analog computers represent mathematical procedures by physical processes, scientific laws lead to representations of physical processes by (almost always) mathematical procedures. (More accurately, I should better state that analog computers *represented* mathematical procedures by physical processes, since such computers were most commonly used prior to the development of digital computers. Digital computers are now so powerful, cheap, and ubiquitous, that I doubt analog computers are used at all any more.)

This antianalog computer is equipped with a set of terminals into which input may be fed and out of which output may be obtained. Different subsets of terminals might be used to feed and obtain different types of input and output. Some terminals might be exclusively either for input or for output, while others might serve for both input and output (though not during the same application). Every input uniquely determines an output. (It is possible, of course, that the same output might be the result of different inputs.) That gives us reproducibility and predictability. A scientific law must also supply a set of rules for translating a physical situation into input acceptable by the antianalog computer (and even for deciding whether a physical situation can be so translated) and for translating the output into physical terms.

As an example, consider the laws of classical vacuum electromagnetism with Maxwell's equations (James Clerk Maxwell, Scottish physicist, 1831–1879) and certain boundary conditions as their antianalog computer. The input might be functions, as a mathematical translation of a physical situation in terms of electric charges and currents. The output is then functions, which are translated as electric and magnetic field strengths, giving the forces on test charges.

As another example, consider the theory of nonrelativistic quantum mechanics with the Schrödinger equation and certain boundary conditions as its antianalog computer. Our input often consists of functions, as a mathematical translation of a physical situation in terms of

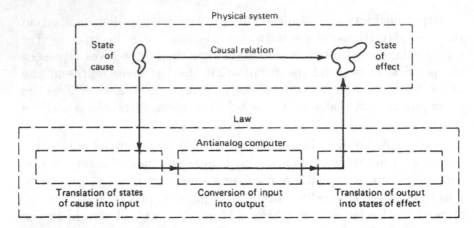

Fig. 4.1. Relation of a law to cause and effect in a physical system

potential and incident wave function. The output is then a function, which is translated as the scattered wave function and from which are derived probability, cross-sections, and such measurable quantities.

Consider a physical system and a law of its behavior. A system must possess at least one pair of subsystems in causal relation for a law of its behavior to be possible, since, as stated above, a law is an expression of a causal relation in a system. Consider any such pair, a cause subsystem and an effect subsystem. Since the law describes the behavior of the whole system, it must take the translation of the states of the cause subsystem as input to its antianalog computer, and the translation of the output of its antianalog computer must, by exhausting its total information content, determine states of the effect subsystem. Moreover, the state of the effect that is causally correlated with a given state of the cause must be consistent with the translation of the output that is uniquely produced by the antianalog computer as a result of receiving as input the translation of the given state of the cause (see Fig. 4.1).

States of causes and effects are actual physical situations occurring in actual physical systems. Each such state is unique; no two are identical. Although similar in many respects, a pair of states might still differ, for example, in their geographical locations, in their instants of occurrence, or in the experimental apparatuses with which they are concerned. In fact, since no physical system is perfectly isolated from the rest of the Universe,[1] in order to make sure we are not missing any-

---

[1] See Sect. 2.2.2. Even if a physical system were isolated from the rest of the Universe with respect to all the known forces and interactions, there would still

thing, we should, to be precise, take the whole Universe as the only physical system worth investigating. Otherwise we run the danger of neglecting some influence, say from the Crab Nebula, that might be crucial. Then states of cause and effect subsystems would be characterized by an infinite number of parameters; we would have to take into consideration the state of every object, field, etc., in the Universe – and at all times too – since they might possibly influence the states of subsystems.

That holistic point of view, with its emphasis on the uniqueness of situations, is unconducive to science. (See the discussion in Sect. 2.2.) In its light everything appears so awesomely complicated that there seems to be no hope of understanding (in the sense of science) anything. Reproducibility and predictability seem utterly beyond reach. Science begins to be possible only when some order is discerned within the confusion, when similarities are remarked upon in spite of the differences. To do science, one must, by decision, guesswork, or blissful ignorance, make assumptions about which influences are more or less important and on those assumptions 'slice' the world into quasi-isolated components. One must then investigate those relatively simple physical systems as if they were really isolated. The simplest systems must be investigated first. Then more complicated systems can be attacked by considering them as syntheses of simpler systems in interaction. This very rough picture gives an idea of the attitude necessary for science to be possible.

Let me add that we have here, it seems to me, the main reason why science, and especially its predictability aspect, developed in the West and not in the East. While the dominant Eastern philosophies emphasized the oneness and wholeness of everything, making science extremely unlikely, if not altogether impossible, the Western weltansicht encouraged analysis, making science most likely, if not inevitable. The different conceptions of the position of *Homo sapiens* vis-à-vis the rest of nature, tying in with religious considerations, were part of the whole scene. But let us return to our business.

We are thus led to the recognition that all scientific laws are doomed in principle to a sort of imprecision. The reasoning is that, if we took

---

remain that influence which, according to the Mach principle (Ernst Mach, Austrian physicist, psychologist, and philosopher, 1838–1916), is the origin of inertia [22]. It would indeed be a great day if we were to succeed in abolishing inertia. In addition, the effect of quantum entanglement is also an anti-isolatory factor, as quasi-isolated systems might be correlated with their surroundings via quantum correlations.

everything into consideration, we would be dealing solely with unique situations, reproducibility and predictability would be impossible, no order could be discerned within the confusion, and science would be inconceivable. It is only by ignoring certain aspects of physical reality that we can discover similarities, obtain reproducibility and predictability, discern order, and find laws.

That essential ignoring is usually taken to be of two kinds. There is *ignoring in practice*, where we admit there exists or might exist influence, but we assume it to be sufficiently weak that it is negligible compared with the effects under consideration. And there is *ignoring in principle*, where we assume there is no influence at all. An example of ignoring in practice is the influence of the star Sirius on our laboratory experiments. We know there are gravitational and electromagnetic influences, but we ignore them, because we know they are negligible. We do not know whether there are other influences, but we assume that, if there are, they too are negligible. An example of ignoring in principle is the influence of 'absolute' position on our experiments; we assume there is none. We assume the laws of nature are the same everywhere in the Universe (refer to Sect. 3.1).

From the empirical point of view it is impossible to distinguish between precise lack of influence and sufficiently weak influence. Experimental investigations can only supply upper bounds for the strength of the effect. Indeed, it is possible that *in principle* there is no precise lack of influence and that all ignoring is ignoring in practice. (Such matters and others are discussed in [23].) *In practice*, however, it is convenient to think in terms of the two kinds of ignoring.

Return now to a law of behavior for a physical system in terms of antianalog computer and translation rules, as described previously. We have just convinced ourselves that the translation of states of a cause subsystem is essentially 'weak' in the sense that certain aspects of the states are ignored in the antianalog computer. And similarly, the output does not contain sufficient information for unique determination of a state of the effect, so that more than one state of the effect is consistent with the translation of an output. That defines equivalent relations for the sets of states of the cause and effect subsystems, according to which states of the cause are equivalent if and only if they translate into the same input, and states of the effect are equivalent if and only if they are consistent with the translation of the same output.

For example, if a law is consistent with the special theory of relativity, all states differing only by spatial or temporal displacements, rotations, or boosts (velocity changes) up to the speed of light are equivalent. States that differ only in this way translate into the same input or are translated from the same output for a relativistic law. In other words, a relativistic law may take interest in any property of a state that is not a position, instant, orientation, or velocity connected with that state.

Or, if a law for a system of particles does not distinguish individual particles, all states differing only by interchange of particles are equivalent.

By that definition of equivalence for the set of states of a cause subsystem, all states belonging to the same equivalence class translate into the same input. This input uniquely determines an output according to the law. And all states of the effect subsystem that are correlated with those states of the cause by the causal relation are translated from the output and are therefore equivalent to each other by definition. It is possible that also states of the effect that are correlated with states of the cause belonging to a different equivalence class, i.e., that translate into a different input, are equivalent to these, since different states of a cause may be causally related to the same state of an effect (but not vice versa) and since different inputs of a theory may determine the same output (while the same input always determines the same output).

Figure 4.2 should help make that clear. Note in the example of the figure that the set of states of the cause decomposes into three equivalence classes, which translate into three different inputs. The antianalog computer converts two of the inputs into the same output and the third input into a different output. The two outputs translate into states of the effect, which decompose accordingly into two equivalence classes.

What we have just proved is the equivalence principle:

Equivalent causes – equivalent effects.

This formulation is relatively imprecise to avoid putting off those who have not yet studied its significance. The intention of the word 'causes' is, of course, states of a cause, and the word 'effects' similarly stands for states of an effect of the cause. The precise formulation is, therefore:

Equivalent states of a cause – equivalent states of its effect.

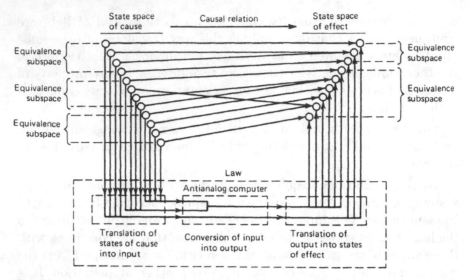

Fig. 4.2. A law imposes equivalence relations on sets of states of cause and effect in physical systems, leading to the equivalence principle

And if we desire to put even more emphasis on the causal relation, we might insert an arrow:

Equivalent states of a cause $\longrightarrow$ *equivalent states of its effect.*

I should mention that the converse of the equivalence principle does not hold as a principle, since equivalent states of an effect are not necessarily causally correlated with equivalent states of a cause, as we saw.

We derived the equivalence principle as a direct consequence of the existence of science. Thus the existence of science requires the existence of the equivalence principle, which is therefore a necessary condition for the existence of science. No equivalence principle (and thus no reproducibility) $\longrightarrow$ no science!

The equivalence principle will be empty if and only if the equivalence relation that a law imposes on states of a cause is the trivial one, that every state is equivalent only to itself. Such a law does not allow reproducibility, so it is not suitable for science. Thus, it is the 'weaknesses', or 'impotencies', of scientific laws, their essential inability to distinguish among the members of certain sets of states, that make the equivalence principle nontrivial (see also Sects. 3.1 and 3.2).

Let us summarize our line of reasoning that led to the equivalence principle. We started with the fact the experimentally reproducible phenomena exist in nature and that causal relations are found in nature. We assumed the validity of predictability, i.e., we assumed that the human intellect is capable of exploiting reproducibility to attain sufficient understanding of natural causal relations to enable prediction of as yet unobserved phenomena. The conceptual tool for the expression of causal relations in physical systems is the law. Thus, our assumption is that the human intellect is capable of inventing laws that describe the behavior of physical systems. This assumption seems to be well justified. A law, by its essential 'impotence', creates an equivalence relation for the set of states of a cause and for the set of states of an effect: States of a cause are equivalent if and only if they translate into the same input, and states of an effect are equivalent if and only if they are translated from the same output. From that and from the character of any causal relation we obtained the equivalence principle:

Equivalent states of a cause $\longrightarrow$ *equivalent states of its effect.*

Or in less precise language:

Equivalent causes – equivalent effects.

4.4 The Symmetry Principle

We are finally in a position to reach our goal of deriving the *symmetry principle*. Although the equivalence principle is fundamental to the application of symmetry in science, in practice it is the symmetry principle, derived directly from the equivalence principle, that is usually used. Its precise formulation, in the language of group theory (see Chaps. 8 and 10), is this:

The symmetry group of the cause is a subgroup of the symmetry group of the effect.

Equivalently:

A symmetry transformation of the cause is also a symmetry transformation of the effect.

Or, expressed in a less technical sounding form:

The effect is at least as symmetric as the cause.

The symmetry principle is also called Curie's principle [24].

To prove the symmetry principle and see what it means, we first define the terms used in it. Consider again a physical system and a law of its behavior. Any state u of the system implies a state for each of its subsystems and, in particular, implies state u_c for a given cause subsystem and state u_e for a given effect subsystem. The law imposes an equivalence relation for the set of states of the cause subsystem (giving the 'equivalent causes' in the equivalence principle) and an equivalence relation for the set of states of the effect subsystem (giving the 'equivalent effects' in the equivalence principle), as discussed in Sect. 4.3.

Define cause equivalence, denoted $\overset{c}{\equiv}$, for the set of states of the whole system as follows: Two states of the whole system are cause equivalent if and only if the states of the cause subsystem implied by them are equivalent. Symbolically,

$$u_c \equiv v_c \quad \Longleftrightarrow \quad u \overset{c}{\equiv} v ,$$

for states u and v of the whole system. Similarly, define effect equivalence, denoted $\overset{e}{\equiv}$, for states of the whole system: Two states of the whole system are effect equivalent if and only if the states of the effect subsystem implied by them are equivalent:

$$u_e \equiv v_e \quad \Longleftrightarrow \quad u \overset{e}{\equiv} v ,$$

for states u and v of the whole system (see Fig. 4.3).

The equivalence principle states that if the states of the cause subsystem implied by states of the whole system are equivalent, then the states of the effect subsystem implied by the same states of the whole system are also equivalent. Symbolically,

$$u_c \equiv v_c \quad \Longrightarrow \quad u_e \equiv v_e ,$$

for states u and v of the whole system. From our definitions we conclude that cause equivalence implies effect equivalence for the set of states of the whole system. Or, symbolically,

$$u \overset{c}{\equiv} v \quad \Longrightarrow \quad u \overset{e}{\equiv} v ,$$

for states u and v of the whole system (see Fig. 4.4).

Consider transformations of states of the whole system. They are changes that carry states of the system into states of the system. Consider in particular those transformations that carry every state into a cause equivalent state. For them any state and its resulting state

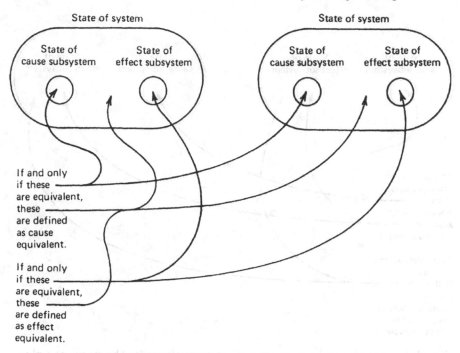

Fig. 4.3. Definition of cause equivalence and effect equivalence for the states of a physical system

imply states of the cause subsystem that are equivalent. Another way of looking at this is in terms of cause equivalence classes. Cause equivalence for the set of states of the whole system decomposes the set into cause equivalence classes. All the states that belong to the same class imply states of the cause subsystem that are equivalent. Now consider those transformations whose effect is to change states of the whole system to states in such a way that any state and its resulting state are within the same cause equivalence class. In other words, such transformations stir up the set of states, but do their stirring only within the cause equivalence classes and do not mix classes. Those transformations are symmetry transformations, symmetry transformations of the cause; they bring about change, but keep intact membership in cause equivalence classes. The family of all symmetry transformations of the cause is called the symmetry group of the cause.

Similarly, transformations that preserve membership in effect equivalence classes are symmetry transformations of the effect, and the family of all those transformations is the symmetry group of the effect.

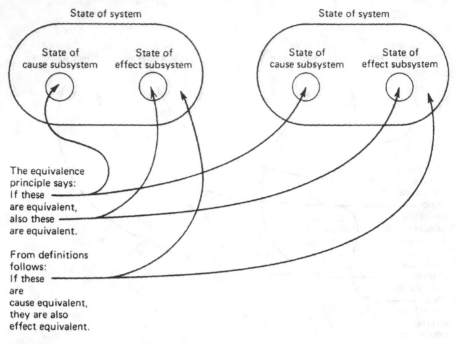

Fig. 4.4. Cause equivalence implies effect equivalence for the states of a physical system

Since, by the equivalence principle and our definitions, cause equivalence implies effect equivalence, it follows that every transformation that preserves cause equivalence must also preserve effect equivalence, that every symmetry transformation of the cause must also be a symmetry transformation of the effect. However, there may be symmetry transformations of the effect that are not symmetry transformations of the cause, since effect equivalence does not imply cause equivalence. That gives us one formulation of the symmetry principle:

> *A symmetry transformation of the cause is also a symmetry transformation of the effect.*

Thus, the symmetry group of the effect contains all the transformations that constitute the symmetry group of the cause and possibly more. When groups stand in such a relationship, the group that is included is called a subgroup of the including group (see Sect. 8.5). That leads to an equivalent statement of the symmetry principle:

> *The symmetry group of the cause is a subgroup of the symmetry group of the effect.*

Degree of symmetry can be quantified by the number of corresponding symmetry transformations (see Sect. 10.7 for more detail). In that sense the symmetry principle can be given the form:

The effect is at least as symmetric as the cause.

Beyond the technical language, what the symmetry principle is telling us is this:

Any symmetry of a cause must appear in its effect. (However, the effect may possess symmetry that is not symmetry of the cause.)

We will discuss applications of the symmetry principle in Chap. 5. But let us now look at one simple example to help see how the symmetry principle works. So imagine we have an electric charge distribution possessing spherical symmetry. This means that any rotation about any axis through the center of the charge distribution leaves the distribution unchanged. (Also, although we do not need it for the present problem, reflection through any plane containing the distribution's center leaves the distribution unchanged.) In addition, imagine that a resting test charge (a point particle carrying a small electric charge) is situated at some distance from the center of the charge distribution. We will use the symmetry principle to find the direction of acceleration of the test charge due to the charge distribution.

What we are actually doing is finding the direction of the electric field of a spherically symmetric charge distribution. But I prefer to formulate the problem in spatiotemporal terms, in terms of acceleration, rather than in terms of the electric field, since the transformation properties of the electromagnetic field are not obvious and I prefer not to get involved with them here. Note that I am also avoiding formulating the problem in terms of force, since, although no harm would have been done, I still feel we are standing on somewhat firmer ground, conceptually, when we assign transformation properties to acceleration rather than to force. Also note that this example is equally applicable to any central force field, not just the electric field, as no use will be made of Coulomb's law (Charles Augustin de Coulomb, French physicist, 1736–1806) specifically.

The cause in this example consists of the charge distribution and the test charge. The effect is the acceleration of the test charge (refer to Fig. 4.5). Any rotation about the axis passing through the test charge and the center of the charge distribution (the dashed line in the

Fig. 4.5. System consisting of a spherically symmetric charge distribution and a resting test charge, as cause, and the acceleration *a* of the test charge, as effect

figure) is a symmetry transformation of the cause. By the symmetry principle, those transformations must be symmetry transformations also of the effect (which might, moreover, possess additional symmetry transformations). Since any nonradial component the acceleration might have would not be carried into itself by those rotations, the acceleration must therefore be radial, i.e., directed precisely toward or away from the center of the charge distribution. Only for these directions are the symmetry transformations of the cause also symmetry transformations of the effect, as required by the symmetry principle. Whether the acceleration is directed toward the center of the charge distribution or away from it depends on the signs of the test charge and the charge distribution and cannot be determined by symmetry considerations [25].

In this section we have been playing free and loose with technical terms that possess rigorous and precise definitions: transformation, group, subgroup, symmetry transformation, and symmetry group. They will be treated rigorously in Chaps. 8–10. What I am attempting to do here is to present and clarify the concepts with only the unavoidable minimum of mathematics. For a fuller understanding, however, the mathematics is needed and is presented with sufficient rigor in Chaps. 8–10.

4.5 Cause and Effect in Quantum Systems

This section is intended for readers with a good background in quantum theory and its Hilbert space formulation.

The application of symmetry in quantum systems is straightforward and successful. It is, in fact, one of the greater success stories of theoretical physics. That is due to the special nature of sets of quantum states, called quantum state spaces, specifically to their being linear vector spaces. This makes it possible to work with realizations of transformation and symmetry groups by groups of matrices acting on the components of the elements (vectors) of quantum state spaces. Such realizations are called linear representations, or representations. The detailed study of the application of symmetry in quantum systems, which is beyond the scope of this book, is well presented in a considerable number of more advanced texts.

It is worthwhile, however, to devote a brief discussion to causes and effects in quantum systems. Causes and effects, as defined previously, are subsystems of the systems under consideration, and subsystems of quantum systems are themselves quantum systems. Thus, causes and effects in quantum systems are quantum systems with all the consequences thereof. For example, in a scattering experiment, if the sharply peaked momentum distribution in the incident beam is part of the cause, one may not expect that the positions of the individual particles in the beam might also be included in the cause, since the latter are in principle indeterminate in the given setup. As for the effect, neither particle trajectories nor scattering angles of individual particles can be part of it. The effect is, in fact, the wave function, especially the 'scattered wave function', from which probabilities and differential cross-sections can be calculated.

If that sounds like the Rutherford experiment, you are quite right, and we should compare the classical and quantum theories of such scattering with regard to cause and effect in each kind of theory. In the classical approach the cause could in principle be the initial position, velocity, and orientation of each incident alpha particle and the position and orientation of each target nucleus. The effect would then be the point where each scattered alpha particle hits the fluorescent detection screen and causes it to flash. However, such a cause is not practically realizable (and is, of course, precluded by quantum principles), and we must make do instead with the statistical properties of the incident beam and the target material. The effect is then correspondingly reduced to statistical properties of the scattered alpha particles.

In nonrelativistic quantum mechanics, on the other hand, the cause is the incident wave function and the scattering potential, which indeed involve statistical properties of the incident and target particles but are different from the classical cause. For instance, the incident wave function has a nonclassical phase (over which we admittedly have no control, but which could be of importance in its relation to other phases in the system). The effect is the scattered wave function, which also contains a phase in addition to statistical information about the scattered particles.

The vacuum state of quantum theory is also due for some discussion. Invariably, even in the most fundamental theories, the properties of the vacuum state are part of the assumptions of the theory, rather than derived from other assumptions. Thus, the vacuum state is part of the cause. That must be kept in mind, especially when the vacuum state is assumed to possess a lower degree of symmetry than the rest of the cause. It is the symmetry of the total cause, including the vacuum state, that, by the symmetry principle, must appear in the effect. Since the vacuum state is crucial in determining the properties of the physical states, as the effect, it should come as no surprise that the physical states are not as symmetric as the cause without the vacuum state, and the situation is in no way a violation of the symmetry principle.

I brought up that point because it has often been presented as an apparent violation of the symmetry principle. The symmetry principle has nothing to fear from such or other *apparent* violations. It has been contrived so that, as long as we are concerned with conventional science (reproducibility, predictability, laws, etc.), it is inviolable.

4.6 Summary

In this chapter we derived the symmetry principle, which is the fundamental principle in the application of symmetry considerations to problem solving in science and engineering and devising theories in physics. We started by discussing in Sect. 4.1 the concept of causal relation in physical systems, whereby certain correlations exist between states of cause subsystems and states of effect subsystems, correlations resulting from the fact that states of subsystems are determined by the states of the whole system.

In Sect. 4.2 we acquainted ourselves with the concepts of equivalence relation and equivalence class.

In Sect. 4.3 we looked into scientific laws as expressions of causal relations. We saw that such laws must ignore certain aspects of states of physical systems. That introduces equivalence relations in the sets of states of systems, from which follows the equivalence principle: Equivalent states of a cause → equivalent states of its effect. From the equivalence principle we derived in Sect. 4.4 the symmetry principle: A symmetry transformation of the cause is also a symmetry transformation of the effect. Equivalently: The symmetry group of the cause is a subgroup of the symmetry group of the effect. Alternatively: The effect is at least as symmetric as the cause. What the symmetry principle means is that any symmetry of a cause must appear in its effect, while the effect may possess symmetry that is not symmetry of the cause.

In Sect. 4.5 we briefly discussed causes and effects in quantum systems.

5

Application of Symmetry

The symmetry principle can be used to set a lower bound on the symmetry of an effect or to set an upper bound on the symmetry of a cause. The former use, which we call the minimalistic use, is characteristic of most practical, technological, and textbook problems, where the law and cause are known for a given system and it is desired to find some or all of the effect, on whose symmetry the symmetry principle sets a lower bound. The problems of basic research, on the other hand, are usually opposite to those. Here one is working from the effect to find the cause, and the symmetry principle sets an upper bound on the cause's symmetry. This is the maximalistic use of the symmetry principle. We start with a discussion of the symmetry principle's minimalistic use.

5.1 Minimalistic Use of the Symmetry Principle

The symmetry principle is used minimalistically in the application of symmetry to problems in which the law and cause are known for a given system and it is desired to find some or all of the effect. Such problems are characteristic of practical, technological, and pedagogical applications. If the cause is known, its symmetry can be worked out. The symmetry principle then states that the effect also possesses this symmetry (and possibly more). The knowledge of the minimal symmetry of the effect is often sufficient for solving the problem fully or partially or at least for simplifying it to some extent. The minimalistic use of the symmetry principle is exploited in varying degrees of sophistication. Here we offer some examples of more simple-minded applications, in order to keep the symmetry considerations in the foreground.

Our first example of minimalistic application of the symmetry principle was presented in Sect. 4.4 of the preceding chapter. There we were given a spherically symmetric electric charge distribution from whose center a resting test charge is situated at some distance. We asked for the direction of the test charge's acceleration. The charge distribution and test charge serve as the cause, with the test charge's acceleration as effect. Applying the symmetry principal minimalistically, we know that any symmetry transformation of the cause must also be a symmetry transformation of the effect. Any rotation about the axis connecting the center of the charge distribution with the test charge is a symmetry transformation of the cause. The requirement that all such rotations be symmetry transformations also of the effect, the acceleration, strongly constrains the direction of the acceleration to the radial, i.e., pointing directly toward or away from the center of the charge distribution.

In the next example we find the direction of the acceleration of a test charge moving parallel to a straight, infinitely long, current-carrying wire. Although the force on the test charge is magnetic, that fact will not enter our discussion. And I purposely formulated the problem in terms of the acceleration of a test charge rather than in terms of the magnetic field around the wire, in order to avoid the complicating issue of the transformation properties of the magnetic field under reflection, a common source of trouble to the unwary.

The cause consists of the current and the moving test charge. The effect is the test charge's acceleration (refer to Fig. 5.1). (Although in the figure the current is given the same sense as the velocity of the test charge, that is immaterial, and the opposite sense serves as well.) The cause is symmetric under the transformation of reflection through the plane containing the wire and the test charge, the plane

Fig. 5.1. System consisting of an infinite straight current and a test charge moving parallel to it, as cause, and the acceleration a of the test charge, as effect

of the page in the figure. The symmetry principle declares that this transformation must also be a symmetry transformation of the effect, so the acceleration can possess no component perpendicular to the plane of the page. Thus, thanks to the symmetry principle, we now know that the direction of the acceleration is confined to the plane. But that is still too loose, and we need to constrain the acceleration even more. The way to do that is to find another symmetry transformation of the cause that, when applied to the effect, removes some or all of the remaining ambiguity.

Such a symmetry transformation is supplied by the compound transformation consisting of the consecutive application of temporal inversion (or time reversal) and spatial reflection (plane reflection) through the plane perpendicular to the current and containing the test charge (at any instant) (see Fig. 5.2). The temporal-inversion transformation by itself reverses the senses of all velocities, in our example both that of the test charge and that of the moving charges that form the current in the wire. Since both the current and the test charge's velocity are perpendicular to the reflection plane of the second part of the compound transformation, their senses are reversed again by the reflection. Thus, the compound transformation indeed leaves the cause intact.

But how does it affect the effect, the acceleration, of which it must be a symmetry transformation, according to the symmetry principle? Temporal inversion by itself does not affect acceleration, because acceleration is the *second* derivative of position. In other words, acceleration is the time rate of change of velocity, and velocity is the time rate of change of position. Thus, acceleration involves two time rates of change, one the rate of change of the other. Under temporal inversion every time rate of change changes sign. That reverses the direction of velocity. But the sign of acceleration is changed twice, so acceleration does not change under temporal inversion. The reflection transformation reverses the component of acceleration perpendicular to the reflection plane. So for the compound transformation to be a symmetry transformation of the acceleration, the acceleration may not possess a component that is perpendicular to the reflection plane. Since neither may the acceleration have a component perpendicular to the page, as we saw earlier, that leaves the acceleration perpendicular to the wire and pointing either toward or away from the wire. Whether the acceleration is directed toward the wire or away from it depends on the sign of the test charge and whether its velocity is in the same sense as the current or in the opposite sense.

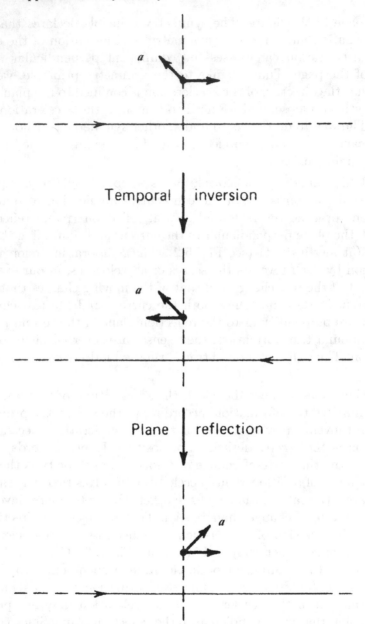

Fig. 5.2. The compound transformation consisting of the consecutive application of temporal inversion and plane reflection is the symmetry transformation of the cause for the system of Fig. 5.1. Its effect on the acceleration *a* of the test charge is shown

In our next example we prove that the orbit of a planet about its sun lies completely in a plane and that this plane contains the center of the sun. Our only assumption is that the sun and planet are each spherically symmetric. Recall that spherical symmetry means the body is symmetric under any rotation about any axis containing its center and under reflection through any plane containing its center. We assume nothing about the nature of the force between the two. (The reasoning will be valid also for a spherically symmetric body in any central force field.)

Consider the situation at any instant. The planet has a certain position and a certain instantaneous velocity relative to the sun. Of all the planes containing the line connecting the centers of the planet and the sun, only one is parallel to the direction of the planet's instantaneous velocity. We call it the plane of symmetry (see Fig. 5.3). The cause consists of the sun and the moving planet. The effect is the planet's acceleration.

The cause has reflection symmetry with respect to the plane of symmetry: The sun and planet are each reflection symmetric with respect to any plane containing their centers, since they are spherically symmetric, and the plane of symmetry contains both centers. The positions of the planet and sun are not changed by reflection through the plane of symmetry, again since the plane contains both their centers. The direction of the planet's instantaneous velocity is parallel to the plane of symmetry, so that velocity is also invariant under reflection

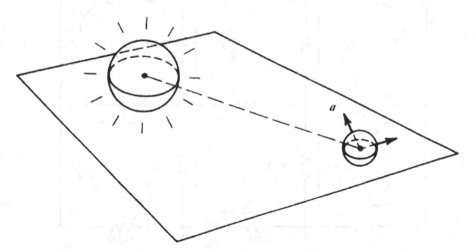

Fig. 5.3. The system consisting of a sun and a moving planet, as cause, and the acceleration a of the planet, as effect. The plane of symmetry is indicated

through the plane. Thus, the cause is reflection symmetric with respect to the plane of symmetry. The effect, the acceleration of the planet, must then possess that symmetry, by the symmetry principle. There-fore, the direction of the planet's acceleration must be parallel to the plane of symmetry. Otherwise, if the acceleration had a component perpendicular to the plane, it and its reflection would not coincide. So the planet's velocity, which is parallel to the plane of symmetry, un-dergoes a change (due to the acceleration) that is parallel to the plane of symmetry (since the acceleration is parallel to the plane) and thus remains parallel to the plane of symmetry. In this way we see that the planes of symmetry of the system at all instants are in fact one and the same plane and that the motion of the planet is confined to that plane.

Our next example of application of the symmetry principle con-cerns electric currents. Consider the DC circuit of Fig. 5.4, certainly good for nothing but an exercise. The six emf (voltage) sources, the 11 resistors, and their connections constitute the cause. The effect con-sists of the resulting currents in the thirteen branches of the circuit and the potential differences between all pairs of points of the circuit. We follow Kirchhoff's rules (Gustav Robert Kirchhoff, German physi-cist, 1824–1887) for finding the currents. By Kirchhoff's junction rule

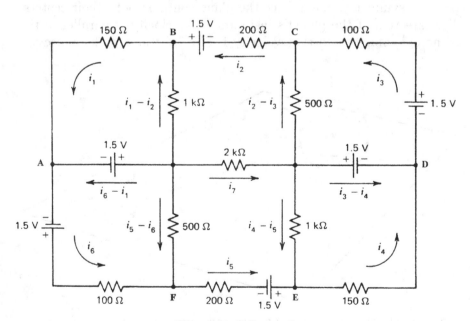

Fig. 5.4. DC circuit

we arbitrarily designate seven of the currents i_1, i_2, i_3, i_4, i_5, i_6, i_7 and express the other six currents in terms of them, as in the figure. Kirchhoff's loop rule then gives us a set of seven simultaneous linear equations for the seven unknown currents, which we do not show here. The equations can be solved and the solutions used to calculate the other six currents. With all the currents known, the potential difference between any pair of points can be calculated with the help of Ohm's law (Georg Simon Ohm, German physicist, 1787–1854). For example, referring to the figure, we might want to know the potential difference between any pair of the points marked A, B, C, D, E, F, such as V_{AB}, V_{AD}, or V_{EF}.

A glance at Fig. 5.4 reveals that the cause possesses symmetry, two-fold rotation symmetry, i.e., symmetry under rotation by 180°, about the axis through the 2-kΩ resistor and perpendicular to the plane of the page. So the symmetry principle can be invoked, and the effect, the currents and potential differences, must have the same symmetry. To help see what that implies, we rotate the system by 180° with the result shown in Fig. 5.5. Now compare the rotated system of Fig. 5.5 with the unrotated system of Fig. 5.4. Symmetry of the effect gives us for the currents $i_1 = i_4$, $i_2 = i_5$, $i_3 = i_6$, $i_7 = -i_7$, so that $i_7 = 0$. For the potential differences symmetry of the effect gives $V_{AD} = V_{DA}$, $V_{BE} =$

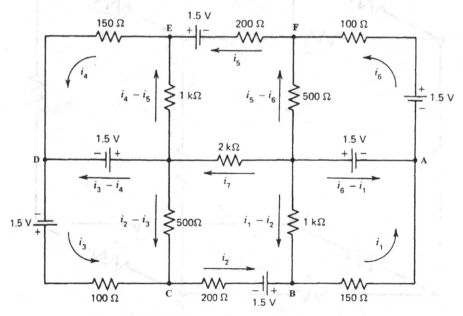

Fig. 5.5. DC circuit for Fig. 5.4 rotated by 180°

V_{EB}, $V_{\mathrm{CF}} = V_{\mathrm{FC}}$. And since by the nature of potential differences $V_{\mathrm{DA}} = -V_{\mathrm{AD}}$, etc., we have $V_{\mathrm{AD}} = V_{\mathrm{BE}} = V_{\mathrm{CF}} = 0$. Symmetry of the effect also gives us relations among the potential differences, such as $V_{\mathrm{AF}} = V_{\mathrm{DC}}$, $V_{\mathrm{CE}} = V_{\mathrm{FB}}$, etc.

Thus, the symmetry principle gives more than half the solution to the problem in this example. Instead of solving seven simultaneous equations for seven unknown currents, we need to solve only three equations for three currents. Three of the desired potential differences are now known, and only half of the others need to be calculated.

Our next example again involves electricity. In this example the system has a higher degree of symmetry than in the preceding one, and the symmetry principle makes the complete solution quite simple, although its explanation is somewhat drawn out. We solve the well known problem of finding the resistance of a network of 12 equal resistors connected so that each resistor lies along one edge of a cube, where the resistance is measured between diagonally opposite vertices of the cube, such as between vertices A and H in Fig. 5.6. Let r de-

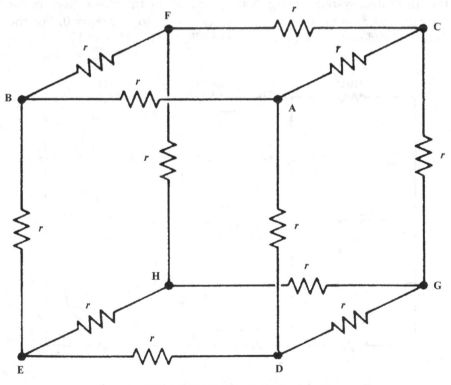

Fig. 5.6. Cube of resistors

note the resistance of each resistor. We imagine a voltage V applied to vertices A and H, as a result of which current i enters the network at A, branches through the resistors, and leaves the network at H. The resistance of the network between A and H is then $R = V/i$ by Ohm's law. The cause is the network and the applied voltage, while the currents in the resistors and the corresponding potential drops (another term for potential differences) on the resistors comprise the effect.

The cause in the present example does not possess the full symmetry of the cube, in spite of all the resistors being equal, since vertices A and H are distinguished from the other vertices and from each other, as the current enters the network at A and leaves at H. So the symmetry transformations of the cause consist of only those symmetry transformations of the cube that do not affect vertices A and H: rotations by 120° and 240° about the diagonal AH (i.e., the diagonal AH is

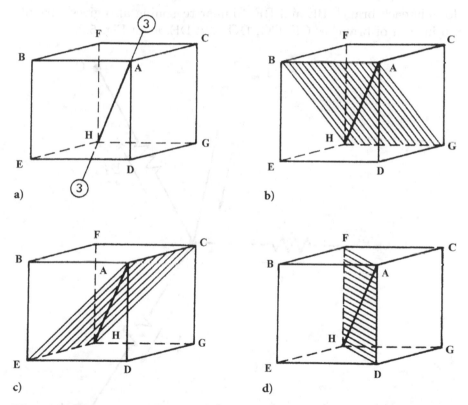

Fig. 5.7. Symmetry of the cause for resistance of the cube of resistors between A and H. (a) Axis of three-fold rotation symmetry. (b)–(d) Planes of reflection symmetry

an axis of three-fold rotation symmetry) and reflections through each of the three planes ABHG, ACHE, ADHF, as in Fig. 5.7. Then by the symmetry principle the effect must also have that symmetry. We can use this fact to find the current in each resistor of the network.

The current i entering the network at vertex A splits among three branches and flows to vertices B, C, and D. Due to the three-fold rotation symmetry, the current divides equally, so that current $i/3$ flows in each branch AB, AC, and AD, as in the diagram of Fig. 5.8. The current $i/3$ entering vertex B then divides again between two branches and flows to vertices E and F. It divides equally between the two branches because of the reflection symmetry with respect to plane ABHG. Thus, current

$$\frac{1}{2}\left(\frac{1}{3}i\right) = \frac{1}{6}i$$

flows in each branch BE and BF. Similar reasoning also gives current $i/6$ in each of branches CF, CG, DG, and DE, as in Fig. 5.9.

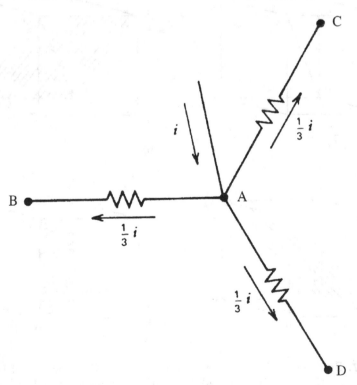

Fig. 5.8. Three-fold rotation symmetry requires equal first division of the current entering the network

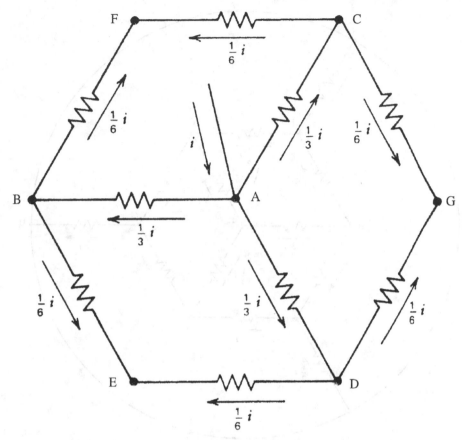

Fig. 5.9. Plane reflection symmetry requires equal second division of the current

After that stage the symmetry takes care of itself. Current $i/6$ enters vertex E from each of vertices B and D, producing current $2 \times i/6 = i/3$ leaving E and flowing to H. Similarly, current $i/3$ enters H from each of vertices F and G. The three currents $i/3$ entering vertex H join to give current $3 \times i/3 = i$ leaving the network, as it should (what enters at A must exit at H). That is illustrated in Fig. 5.10.

Now, the voltage V between A and H equals the sum of potential drops between those two vertices, where the sum may be calculated over any continuous path in the circuit that connects A and H. Let us choose path ABEG. By Ohm's law the potential drop on a resistor equals the product of the current in the resistor and its resistance. So, referring to Fig. 5.10, the potential drop from A to B is $ir/3$, from B to E the potential drop is $ir/6$, and from E to H it is $ir/3$. Adding those

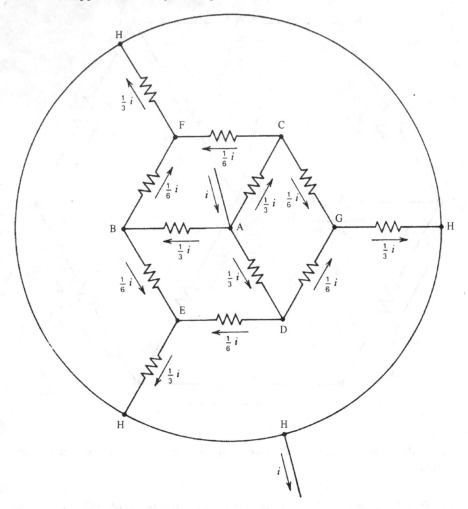

Fig. 5.10. Rejoining of divided currents in two stages before leaving the network

together, we obtain $V = 5ir/6$. Thus, the resistance of the network between A and H is

$$R = \frac{V}{i} = \frac{5ir/6}{i} = \frac{5}{6}r \,.$$

This is the solution of the problem.

In the last two examples note that it is not the actual geometry of the circuits that is important, but rather their electric structure. Thus, it is not the geometric symmetry of the circuits that really interests us, but rather the permutation symmetry of their components, the fact that the circuit remains unchanged under certain interchanges of components. It is true that we expressed that permutation symmetry in geometric terms in order to keep our considerations as familiar as possible. However, the former circuit does not have to be laid out as nicely as in Fig. 5.4, nor does the latter network actually have to lie along the edges of a cube for our symmetry argument to hold. Thus, it is really permutation symmetry that we are using, based on the functional, rather than geometric, equivalence of parts of the circuit. This point is important to keep in mind in the application of symmetry to physical systems. Functional equivalence of parts of a system can be a source of permutation symmetry.

Another point to note in the last two examples is that each problem has a unique solution. So naturally the solution was chosen to be the effect in each case, and application of the symmetry principle was straightforward. Uniqueness of solution is characteristic of passive electric circuits. That is not always the case, however, when circuits contain active components, such as transistors. Then more than one solution might be possible. What is the effect in that case and how is the symmetry principle applied? Read on.

Our next example of minimalistic application of the symmetry principle is of somewhat different character. It is often assumed that the static electric charge distribution on the surface of an isolated charged conducting sphere is homogeneous, i.e., that the surface charge density in such a case possesses spherical symmetry. And if any explanation is offered at all, it is invariably simply a reference to 'symmetry considerations'. Now, it is indeed true that the surface charge density on an isolated charged conducting sphere in the static case possesses spherical symmetry. This can be shown, using the fact that the surface of a conducting sphere is then an equipotential surface and relying on uniqueness considerations. This has not been shown, as far as I know, by proving that the spherically symmetric charge configuration is the configuration of lowest electrostatic potential energy. As for 'symmetry considerations', they certainly do not require a spherically symmetric charge distribution, as we will see, although they do allow it.

How does the sphere get its charge to begin with? We must charge it. We then remove the charging device far enough away so that the sphere can be considered isolated. The cause consists of the charging device and the sphere, and the effect is the final charge distribution on the sphere. The most symmetric practically attainable cause I can think of is attained by keeping the charge source very far from the sphere at all times and connecting the two through a straight wire that touches the sphere perpendicularly to the sphere's surface, so that the line of the wire passes through the center of the sphere. When charging is complete, the wire is withdrawn from the sphere along its line. (Alternatively, a spherical charge source could be put in contact with the sphere and then removed along the line containing the two centers.)

Thus, the most symmetry we can give the cause is axial and reflection symmetry, symmetry under all rotations about the line of the wire (or the line containing the centers) and reflections through all planes containing that line (see Fig. 5.11). This is not spherical symmetry, which is symmetry under all rotations about all axes through the center of the sphere and reflections through all planes containing the center. The latter group of transformations obviously includes the former. Indeed, the sphere itself is spherically symmetric. However, the *total* cause possesses only the axial and reflection symmetry described. So the only symmetry we can demand of the effect, the surface charge distribution, is, by the symmetry principle, also axial and reflection symmetry. Even so, spherical symmetry is not excluded by the symmetry principle, since the effect may be more symmetric than the cause.

If we do not know or prefer to ignore the way the sphere is charged and wish to consider the sphere itself as the cause, the effect will certainly not be the charge distribution, which might depend on the way the sphere is charged, but rather the set of all allowed charge distributions. Since the cause in this case is spherically symmetric, so is the effect, by the symmetry principle. Thus, even though no single allowed charge distribution has to be spherically symmetric, given any one allowed charge distribution, all those obtained from it by any rotation about any axis through the center of the sphere or reflection through any plane containing the center are also allowed charge distributions [26].

That brings us to the following warning. If the state of what is chosen as the cause in a problem is not sufficient to determine a unique

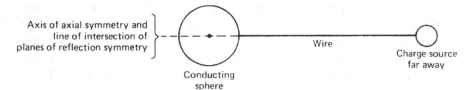

Fig. 5.11. Charging of a conducting sphere

solution to the problem, then the effect cannot be the solution, but is rather the family of all solutions consistent with a state of the cause. For another example of this, note that the symmetry principle is applicable not only to physical systems, but to any system in which a causal relation exists. Such is an equation and its set of solutions. Then, given as the cause an equation possessing some symmetry, the effect is not any particular solution of the equation, and indeed no solution need possess the full symmetry of the equation. The effect is the family of all solutions of the equation, and that is what must have the symmetry of the equation. Given any one solution, the application to it of any symmetry transformation of the equation produces another solution (or even possibly the same solution). If, however, additional conditions are involved, such as boundary or initial conditions or constraints, they may be included in the cause and will usually reduce its symmetry.

To give a more specific example, consider the algebraic equation

$$x^2 - 9 = 0 \, ,$$

or any other polynomial equation involving only either even or odd powers of the unknown variable. Every such equation is symmetric under the transformation

$$x \longrightarrow -x \, .$$

(The odd-power polynomials change sign under the transformation, but the equations are the same, since the equation $-P = 0$ amounts to the same as the equation $P = 0$ for polynomial P.) If each root of such an equation had the same symmetry, the only allowed root would be $x = 0$. However, it is the set of all roots of the equation, and not each root, that must possess the symmetry of the equation, with the result that for equations of the type we are discussing all nonzero roots must appear in positive-negative pairs. For instance, the roots of the equation just presented are $x = \pm 3$.

Or consider the harmonic-oscillator differential equation

$$\frac{d^2 y}{dt^2} + a^2 y = 0 \, , \qquad a = \text{constant} \, .$$

It possesses symmetry under each of the following transformations:

- temporal displacement: $t \longrightarrow t + b$, for arbitrary constant b,
- temporal inversion: $t \longrightarrow -t$,
- spatial dilation: $y \longrightarrow cy$, for arbitrary, positive constant c,
- spatial inversion: $y \longrightarrow -y$.

Since no initial conditions are specified, the family of solutions of this equation must possess the same symmetry as the equation. Thus, if

$$y = f(t)$$

is a solution, so must also

$$y = f(t + b), \qquad y = f(-t), \qquad y = cf(t), \qquad y = -f(t)$$

be solutions for arbitrary allowed constant b and c. For the oscillator equation the family of all solutions can be written in the forms

$$y = A\sin(at + B) = C\sin at + D\cos at ,$$

for all constant A and B or C and D. Under the temporal-displacement transformation the general solution becomes

$$\begin{aligned} y &= A\sin(at + ab + B) \\ &= A\sin(at + B') , \qquad B' = ab + B . \end{aligned}$$

Since B' ranges over the same values as does B, we have recovered the same family.

Under the temporal-inversion transformation the general solution becomes

$$\begin{aligned} y &= A\sin(-at + B) \\ &= A\sin(at - B) \\ &= A'\sin(at + B') , \qquad A' = -A , \qquad B' = -B , \end{aligned}$$

and again we find that the family is symmetric.

Under the spatial-dilation transformation the general solution becomes

$$\begin{aligned} y &= cA\sin(at + B) \\ &= A'\sin(at + B) , \qquad A' = cA , \end{aligned}$$

whence we have symmetry again.

And under the spatial-inversion transformation the general solution becomes

$$y = -A\sin(at + B)$$
$$= A'\sin(at + B) , \qquad A' = -A ,$$

and is symmetric.

If initial conditions are imposed by specifying the values of the solution and its derivative for some value of t, the particular solution will in general not possess any of the above symmetries, although in certain cases it might retain one or more of them [27].

Note that each solution of the oscillator equation possesses the property of periodicity, i.e., it is symmetric under a certain minimal temporal displacement and integer multiples of it,

$$y\left(t + \frac{2\pi}{a}n\right) = y(t) ,$$

for

$$n = 0, \pm 1, \pm 2, \dots .$$

The minimal temporal displacement, or period, is $2\pi/a$. This symmetry does not follow from the symmetry of the equation. It can be discovered only by actually solving the equation. For comparison, the equation

$$\frac{d^2 y}{dt^2} - a^2 y = 0 , \qquad a = \text{constant} ,$$

which is very similar to the oscillator equation, possesses the same four symmetries that we found for that equation, but does not have periodic solutions. Its general solution can be written

$$y = C\sinh at + D\cosh at ,$$

and the hyperbolic functions are not periodic. The family of its solutions must and does possess the four symmetries of the equation.

Finally, to end this section, I will present an apparent violation of the symmetry principle and its resolution. According to Hermann Weyl (German-American mathematician, 1885–1955) [28], Ernst Mach "tells of the intellectual shock he received when he learned as a boy that a magnetic needle is deflected in a certain sense, to the left or to the right, if suspended parallel to a wire through which an electric current is sent in a definite direction." Refer to Fig. 5.12. Let us see why that phenomenon appears to violate the symmetry principle, to Mach's shock. The cause in that case is the electric current in the wire and the magnet hanging above the wire. The effect is the deflection of the magnet. Presumably, as a boy, Mach had not yet learned about the nature of a magnet. So the cause, as he perceived it, was symmetric under reflection through the vertical plane that contains the wire and the magnet. In other words, left and right appeared to be equivalent in the cause, with neither distinguished from the other. By the symmetry principle, then, the deflection of the magnet should possess the same symmetry and should not be either to the left or to the right, which would distinguish one sense from the other. When the magnetic needle was nevertheless deflected in a certain sense (to the left or to the right), Mach could not see what there was in the cause that distinguished that sense from the other sense.

When the symmetry principle appears to be violated, we must question our understanding of the cause and the effect. In this case, it was the nature of a magnet that Mach was not understanding. Simply put, the reflected magnet does not coincide with the original, even if it visually seems to. A magnet derives its magnetism from aligned atomic magnets, where each such magnet is essentially a microscopic electric current loop. Those currents are flowing in a certain sense, either clock-

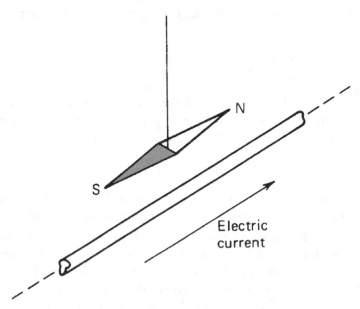

Fig. 5.12. Mach's dilemma

wise or counterclockwise, when viewed along the axis of the magnet, along the line through the magnet's poles. In the example of Fig. 5.12, if we looked along the axis from the south pole (S) to the north pole (N), and if our vision were sufficiently sharp, we would see the atomic current loops flowing clockwise. Clearly, then, reflection through the vertical plane that contains the wire and the magnet does *not* leave the situation unchanged. It changes the clockwise currents to counterclockwise currents and thus *reverses the polarity of the magnet*. Since the cause is not reflection symmetric, neither need the effect, the magnet's deviation, be reflection symmetric, and indeed it is not. In the setup of Fig. 5.12 the north pole of the magnet deflects to the right, as viewed from S to N by a person standing normally.

5.2 Maximalistic Use of the Symmetry Principle

The problems of basic scientific research are usually opposite to those for which the symmetry principle is used minimalistically. In basic research it is the effect that is given, and we try to find the cause. The effect in such a problem is one or more experimental phenomena, and we attempt to discern order among them, find laws for them, and devise a theory explaining those phenomena as being brought about by some cause. A theory is considered to be better the more phenomena it explains and the simpler the cause that is supposed to be producing them [29]. Although simplicity is not a standardized concept and is at least somewhat a matter of taste, it is generally agreed that symmetry contributes greatly to simplicity.

The reason that a symmetric situation is simpler than an asymmetric one is that the former involves less conceptual raw material. For example, in an amorphous solid the atoms (or molecules) are located at more or less random positions and there is no microscopic spatial symmetry. (Viewed *macroscopically*, however, the material is approximately homogeneous and isotropic, i.e., has no distinguished locations or directions. This implies approximate spatial symmetries on the macroscopic scale, but that is another matter.) The microscopic situation is fully described by the specification of the location of every atom, which is a tremendous amount of information. In a crystalline solid, on the other hand, the atoms (or ions or molecules) are located on a lattice, i.e., they are arranged in a spatially periodic array. The situation possesses certain spatial symmetries: definitely spatial-displacement symmetry and possibly also others, such as rotation and reflection symmetries. To describe the situation, it is sufficient to specify the locations of the few atoms composing the unit cell and the repeat distance of the unit cell in three independent directions. Thus, very much less information is involved here than in the amorphous case and in that sense the crystalline situation is simpler than the amorphous one. The symmetry clearly and strongly brings about the simplicity.

For another example, at a more fundamental level, imagine we have a physics theory that involves different laws of nature in different directions. According to such a theory, the same system will evolve differently when it is oriented in different directions. Such a theory does not possess rotation symmetry. Compare that with a rotation symmetric theory, one that gives the same law in all directions. Rather than a different physics in each direction, we now have the physics of a single direction repeated in all directions. The latter is considered to be much simpler than the former.

Thus, in devising a theory for a given set of experimental phenomena, we usually assume as symmetric a cause as possible. And how symmetric can a cause be? Here the symmetry principle serves to set an upper bound on the symmetry of the cause: The cause can be no more symmetric than the effect. That is what we call the maximalistic use of the symmetry principle.

So we must first identify the symmetry of the phenomena we wish to explain. That symmetry is often far from obvious. Then we construct our theory so that the cause will have just the same symmetry, if possible. If it is not possible to assign maximal symmetry to the cause, we must assume a less symmetric cause and include in the theory an explanation of why the effect is more symmetric than the cause. But we may never assume that the cause has more symmetry than do the phenomena being explained, which would violate the symmetry principle.

What most often happens, though, is that the symmetry of a set of phenomena is only approximate. (We discuss approximate symmetry in Sects. 1.1 and 6.1.) In such a case the first step toward a theory is to determine the ideal symmetry that is only approximated by the phenomena. That can be very difficult, if the symmetry is far from exact. Then, to obtain as symmetric a cause as possible, we try to construct a theory such that the cause will contain a dominant part possessing the ideal symmetry of the effect and another, symmetry breaking part that does not have the symmetry. In the (possibly hypothetical) limit of complete absence of symmetry breaking, the dominant part of the cause produces the ideal symmetry of the phenomena, while the symmetry breaking part brings about the deviation from ideal symmetry. A complete theory contains the symmetry breaking mechanism within its framework. But sometimes the cause of the symmetry breaking must be left as a mystery to be cleared up when more experimental facts are known or a better theory can be found.

For a classic example of the maximalistic use of the symmetry principle we turn to nuclear physics. The basic problem of nuclear physics is the strong nuclear interaction, the force that binds protons and neutrons together to form nuclei. This force is not yet completely understood, but appears to be explainable in terms of the interactions among the quarks that compose the proton and neutron. Those interactions are mediated by gluons and are described by the SU(3) color gauge theory (see Sect. 3.6). Protons and neutrons are also affected by the weak nuclear interaction, but at a strength much lower than that of the strong interaction. This interaction is also not yet completely understood, but seems to be described by another gauge theory.

On the other hand, the electromagnetic interaction among protons and neutrons as well as the Pauli exclusion principle (Wolfgang Pauli, Swiss–Austrian physicist, 1900–1958) are very well understood. (The Pauli principle states in the present context that no two protons or no two neutrons can be in the same quantum state.) The effect consists of such phenomena as the various properties of all kinds of nuclei and the results of scattering experiments, in which protons and neutrons are made to collide and nuclei are bombarded with electrons, protons, neutrons, other nuclei, etc. The effect is found to exhibit the following approximate symmetry. Two kinds of nuclei differing only in that one of the neutrons in one kind is replaced by a proton in the other often possess certain similar properties (although their electric charges, for instance, are clearly unequal). Also, in scattering experiments similar results are obtained whether the interacting particles are two protons, two neutrons, or a proton and a neutron.

Thus, nuclear phenomena are approximately symmetric under interchange of proton and neutron. That symmetry is called charge symmetry, since the major difference between the proton and the neutron is their different electric charges. The symmetry principle then suggests that we assume the strong nuclear interaction is exactly charge symmetric, i.e., completely blind to any difference between the proton and the neutron. The symmetry breaking factors are assumed to consist of the electromagnetic interaction, which discriminates between proton (electrically charged) and neutron (electrically neutral), the weak nuclear interaction, which also discriminates between proton and neutron, and the Pauli principle, which discriminates between identical particles and different particles. That assumption has, in fact, proved to be very successful.

Another example of the maximalistic use of the symmetry principle is in the study of the elementary particles and their interactions (of which nuclear physics is a particular case). Here the symmetries of the experimental phenomena are much more complicated and much more approximate (i.e., less nearly exact, except for certain exact symmetries) than in nuclear physics. Maximalistic use of the symmetry principle has guided physicists to an understanding that is called the standard model. This scheme is based on a framework in which the elementary particles of nature are composed of two sets: matter particles and force particles. The set of matter particles in turn comprises two sets: hadrons and leptons. There are four interactions among the matter particles: the strong, weak, electromagnetic, and gravitational interactions. The interactions are mediated by the force particles. For every type of particle there exists a corresponding antiparticle, although some particles are their own antiparticles. The standard model is concerned solely with the strong, weak, and electromagnetic interactions, ignoring gravitation [30–32].

Of the matter particles, the strong interaction affects only the hadrons. The set of hadrons comprises six quarks, which are classified in pairs and given very fanciful names: (up, down), (strange, charmed), (bottom, top). (The nuclear particles of ordinary matter, which are the proton and neutron – collectively called nucleons – are composed of quarks.) Each quark has three states, designated, again fancifully, by the colors red, blue, and green. The strong interaction is mediated by a set of eight gluons. In addition to the strong interaction, the hadrons are subject also to the other three interactions.

The leptons are affected only by the weak, electromagnetic, and gravitational interactions. The set of leptons consists of six particles, which are classified in pairs: (electron, electron neutrino), (muon, muon neutrino), (tau, tau neutrino). (These three pairs are suspected of being related, in some as yet to be understood way, to the three pairs of quarks.) The weak interaction is mediated by three 'intermediate bosons', designated W^+, W^-, Z^0. The mediator of the electromagnetic interaction is the photon, and that of the gravitational interaction is the graviton.

The standard model describes each of the three interactions it is concerned with by a gauge theory, a theory based on gauge symmetry (see Sect. 3.6). The gauge symmetry of the strong interaction is denoted SU(3), where the 3 comes from the number of quark colors. This is symmetry under transformations that mix the colors. For the weak interaction the gauge symmetry is called SU(2), with the 2 referring to the two members of each pair of hadrons and leptons. The transformations involved in this symmetry mix the members of each pair. The gauge symmetry of the electromagnetic interaction is named U(1) and has to do with changes of phase, an abstract quantity that characterizes states of particles. The complete gauge theory of the standard model comprises the three gauge theories. Its symmetry is denoted SU(3)×SU(2)×U(1), indicating the three independent symmetries of the theory.

The standard model does not tell the whole story and leaves much to be explained. Physicists hope that experimental results to be obtained in the near future from new, more powerful particle accelerators will shed light on physics beyond the standard model. Such physics is expected to involve even more symmetry than that of the standard model. One such extension that has already been recognized unifies the weak and electromagnetic interactions into the electroweak interaction. The electroweak gauge theory is a more symmetric theory than the two theories it unifies, since it possesses symmetry transformations in addition to all the symmetry transformations of the separate theories. Such additional transformations might mix the photon and the Z^0, for instance.

A further extension is envisioned, although its character is still unclear. Called grand unified theory (GUT), it would unify the electroweak and strong interactions and should give a good description all non-gravitational phenomena. The GUT gauge theory would possess even more symmetry than do the strong and electroweak theories it unifies. Certain of its symmetry transformations that would not be symmetry transformations of the two separate theories might mix hadrons and leptons, for example.

Even that does not exhaust the capacity of physicists' imaginations. The gravitational interaction, which has so far been left out of the picture (for very good reasons, which we will not discuss here), might be unified with the other three in what is often called a theory of everything (TOE). A TOE would be even more symmetric than any of the theories mentioned above [30, 33].

A different kind of extension of physics beyond the standard model involves what is called supersymmetry (SUSY). This is symmetry under transformations that mix matter particles and force particles. It requires the existence of additional kinds of particle to those that the standard model deals with, particles that have not (yet) been discovered or produced. Evidence relating to SUSY is anticipated in the near future from new particle accelerators [31].

Another approach to unification goes under titles like 'string theory', 'superstring theory', and 'M-theory'. Not only does it attempt to describe nature at a more fundamental level than the above-mentioned theories do, but it requires the existence of additional spatial dimensions to the three we are familiar with. Its fundamental entities, called strings, have sizes of the order of 10^{-35} meter and are assumed to oscillate in various ways and to join and split. This theory has been under development for some time now, but has yet to produce results that are experimentally testable, even by new particle accelerators. See [30, 34], and for an opposing view see [35, 36].

5.3 Summary

This chapter dealt with the application of the symmetry principle. In its minimalistic use, the symmetry principle sets a lower bound on the symmetry of an effect, since, according to the principle, the symmetry of an effect cannot be less than that of its cause. In Sect. 5.1 we examined examples of minimalistic use of the symmetry principle. Some of the examples involved physical systems, while others were mathematical. The physical cases had the character of problem solving for a unique solution, and the solution was fully or partially found by applying the symmetry of the cause to the effect. The mathematical examples served to examine how to deal with situations in which a problem possesses more than one solution. In such cases it is the family of all solutions that serves as the effect and must exhibit the symmetry of the cause (and possibly more).

In Sect. 5.2 we discussed the maximalistic use of the symmetry principle. In that use the principle sets an upper bound on the symmetry of a cause, since, by the symmetry principle, the cause cannot be more symmetric than its effect. Maximalistic use is characteristic of basic science research, in which the effect is given and its cause must be found. For a cause to be as simple as possible, it must be as symmetric as possible, and its maximal allowed symmetry is that of the effect. The examples were from the fields of nuclear physics and the physics of elementary particles and their interactions.

6

Approximate Symmetry, Spontaneous Symmetry Breaking

In this chapter we briefly discuss approximate symmetry, based on the concept of a metric for a set of states of a system. We present the concepts of approximate symmetry transformation, exact symmetry limit, broken symmetry, and symmetry breaking factor. We then more substantially discuss spontaneous symmetry breaking, situations in which the symmetry principle – that the effect is at least as symmetric as the cause – seems to be violated. The essence of the matter is stability of the symmetry of an effect under perturbations of the symmetry of its cause.

6.1 Approximate Symmetry

In Sect. 1.1, I presented a conceptual definition of approximate symmetry. However, in order to express approximate symmetry within the general symmetry framework that we started developing in Chaps. 3 and 4 and will formalize further in Chap. 10, we need the notion of *approximate symmetry transformation*. That is any transformation that changes every state of a system to a state that is nearly equivalent to the original state. And just what does 'nearly equivalent' mean? For that we must soften the all-or-nothing character of the equivalence relation, upon which symmetry is based (see Sects. 4.2–4.4), in order to allow, in addition to equivalence, varying degrees of inequivalence. The way to do that is to find a *metric* for a set of states of a system, a 'distance' between every pair of states, such that null 'distance' indicates equivalence and positive 'distances' represent degrees of inequivalence. In Sect. 10.6 we will work out the formal details of metrics. Here we will stick with concepts.

A metric is an expression of the physical properties of the system. States that are 'close' to each other are 'almost' equivalent in a meaningful physical sense. States that are 'farther' from each other are 'less nearly' equivalent in a physically significant way. Consider the simple example of approximate symmetry in Fig. 1.6 in Sect. 1.1. The figure possesses no exact symmetry; there is no transformation that brings the figure into exact coincidence with itself. Rotation by 180° about the center in the plane of the page brings the figure into a state that nearly coincides with the original state. Rotations by angles that progressively differ from 180° transform the figure into progressively less near coincidence. So a metric for this system would give a small 'distance' between states that differ by 180° rotation and progressively larger 'distances' between states that differ by rotations through angles that progressively deviate from 180°. States that differ by 90° or 270° are the most different from each other, and for such pairs a metric would give the greatest 'distance'. As the angular difference between pairs of states decreases from 90° or increases from 270°, the states become closer to coincidence, and the closer the angular difference is to 0° or to 360°, the more nearly the states coincide. So a metric would give decreasing 'distances' between the members of such pairs, 'distances' that approach zero as the angular difference approaches 0° or 360°. Perfect coincidence occurs for 0° or 360°, but those 'rotations' do nothing, and we have merely the trivial coincidence of every state with itself. Any metric must give null 'distance' between every state and itself.

Now that we have introduced the concept of a metric, we can define an approximate symmetry transformation for a set of states equipped with a metric. It is any transformation that transforms all states into states that are sufficiently nearly equivalent to the originals, where 'nearly equivalent' means within some specified 'distance' as given by the metric. Thus, what is or is not considered an approximate symmetry transformation depends on how great a deviation from equivalence one is willing to tolerate, where that is expressed by the greatest metric 'distance' one will accept.

In the example of Fig. 1.6, that might result in my determination that only rotations in these ranges are approximate symmetry transformations: 175°–185°, 0°–8° (not including 0°), and 352°–360° (not including 360°). The first range consists of rotation angles in the neighborhood of 180° ± 5°. The other two ranges together can be thought of as rotations in the range −8° to +8° (excluding 0°), where a negative rotation is performed in the opposite sense to that of a positive rotation, and the result of rotation by $-a°$ is the same as that of rotation by $360° - a°$. However, you might perhaps be more stringent than I am about what is to be considered nearly equivalent. So for you only rotations in the more limited ranges of $180° \pm 3°$ and $-6°$ to $+6°$ (excluding $0°$) might be considered approximate symmetry transformations.

For a set of states equipped with a metric and a set of approximate symmetry transformations for the set of states and metric, an *exact symmetry limit* is another metric for the same set of states for which some of the approximate symmetry transformations (according to the original metric) are exact symmetry transformations, for which some near equivalences (according to the original metric) become exact equivalences. Since we are considering physical systems, so that the metric has physical significance, an exact symmetry limit of it might correspond to a physically realizable system, to a conceivable but physically unrealizable system, or to nothing conceivable as a physical system at all.

We can use the example of Fig. 1.6 again. Since states that differ by $180°$ are nearly equivalent, so that a metric would give a small 'distance' between such states, a natural exact symmetry limit would be a new metric that is similar to the original one but now gives null 'distance' for such states. The new metric describes a system, similar to the original one, for which pairs of states that differ by $180°$ are exactly equivalent. That is not only conceivable but easily realizable. Simply reduce the size of the lower-right X and arcs to match the upper-left X and arcs (or enlarge the latter to match the former) and – voila! – we have a realization of the exact symmetry limit. For the altered system a rotation by $180°$, which was an approximate symmetry transformation for the original system, is an exact symmetry transformation.

Returning to the nuclear physics example of maximalistic use of the symmetry principle presented in Sect. 5.2, we have charge symmetry as an approximate symmetry. An exact symmetry limit would have charge symmetry as an exact symmetry, i.e., the strong nuclear interaction would be acting, while the weak nuclear interaction, the electromagnetic interaction, and the Pauli exclusion principle would be 'switched off'. Such a situation is conceivable, since by describing it just now we are conceiving of it, but it certainly is not physically realizable. In terms of metrics, any realistic metric for the actual situation must give nonzero 'distances' for pairs of states differing only in that one or more protons in one state are replaced by neutrons in the other and vice versa, since the weak nuclear interaction, the electromagnetic interaction, and the Pauli principle distinguish between such states. An exact symmetry limit for charge symmetry would give null 'distance' for such pairs, expressing their equivalence in the hypothetical absence of the three effects that ruin the exact symmetry.

That brings us to another term for approximate symmetry, which is *broken symmetry*. The *symmetry breaking factor* is whatever factor the (possibly hypothetical) switching off of brings about an exact symmetry limit. In the example of Fig. 1.6 the symmetry breaking factor is the difference in size between the two Xs and pairs of arcs. When that difference is eliminated, the approximate symmetry, or broken symmetry, becomes exact. In the nuclear-physics example the weak nuclear interaction, the electromagnetic interaction, and the Pauli principle constitute the symmetry breaking factor. In their hypothetical absence, approximate charge symmetry, or broken charge symmetry, would be an exact symmetry.

Another example is a crystal, which possesses broken spatial displacement symmetry. The exact symmetry limit is an infinite perfect crystal, obviously unobtainable in practice. The symmetry breaking factor here comprises the real crystal's finiteness and defects. Or, consider the four-equal-straight-armed figure of Fig. 6.1, where each pair of opposite arms forms a straight line segment and the angle δ is almost, but not quite, a right angle. That system possesses approximate

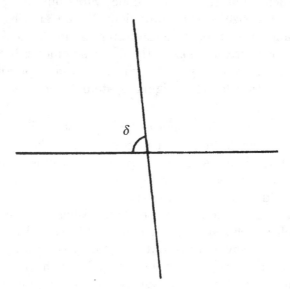

Fig. 6.1. A system possessing approximate four-fold rotation symmetry for $\delta \neq 90°$ but $\delta \approx 90°$

four-fold rotation symmetry. The symmetry breaking factor is the difference between δ and $90°$. The exact symmetry limit is the system with $\delta = 90°$ and is realizable.

6.2 Spontaneous Symmetry Breaking

In discussions of symmetry and the symmetry principle, that the effect is at least as symmetric as the cause, the question of 'spontaneous symmetry breaking' inevitably arises. There appear to be cases of physical systems in which the effect simply has less symmetry than the cause, where the symmetry of the cause is possessed by the effect only as a badly broken symmetry. In those cases the family of exact symmetry transformations of the effect forms part of the family of symmetry transformations of the cause, rather than vice versa as the symmetry principle requires. In other words, we seem to have broken symmetry with no symmetry breaking factor. Among those cases are Garrett Birkhoff's hydrodynamic 'symmetry paradoxes' (American mathematician, 1911–1996) [37], and additional examples are noted later in this section. Yet, as we have convinced ourselves, the symmetry principle cannot but hold.

The trouble is that what is taken to be the cause is not complete, and the complete cause is less symmetric. What is assumed to be the exact symmetry of the cause is really only an approximate symmetry, although a very good approximation. The approximation is so good that we are deceived into believing that it is perfect and overlook the small symmetry breaking factor that makes the symmetry only approximate. What is in fact usually overlooked is the influence of small, random fluctuations in physical systems. Thus, the symmetry principle is saved. The exact symmetry of the cause remains, as it must, the minimal exact symmetry of the effect. But a good approximate symmetry of the cause can be possessed by the effect as a badly broken symmetry.

Just how do small, symmetry breaking perturbations of a cause affect the symmetry of the effect? What can be said about the symmetry of an effect relative to the *approximate* symmetry of its cause? That depends on the actual nature of the particular physical system, on whatever it is that links cause and effect in each case. But we can consider the possibilities:

1. *Stability*. The deviation from the exact symmetry limit of the cause, introduced by the perturbation, is 'damped out', so that the approximate symmetry of the cause is the minimal exact symmetry of

the effect. In this case there is no spontaneous symmetry breaking and the symmetry principle does not appear to be violated.

2. *Lability*. The approximate symmetry of the cause serves as the minimal approximate symmetry of the effect with similar degrees of approximation. Here too the symmetry principle does not appear to be violated.

3. *Instability*. The deviation from the exact symmetry limit of the cause, introduced by the perturbation, is 'amplified', and the minimal symmetry of the effect is only the exact symmetry of the cause (including perturbation), with the approximate symmetry of the cause appearing in the effect as a badly broken symmetry. This is what is commonly called *spontaneous symmetry breaking*. To quote Birkhoff [37]:

> *Although symmetric causes must produce symmetric effects, nearly symmetric causes need not produce nearly symmetric effects*: a symmetry problem need have no *stable* symmetric solutions.

An example of a stable situation is offered by the solar system. All the planetary orbits lie pretty much in the same plane, and that seems to have been the case for as long as observations have been recorded, or even for as long as the planets have existed, according to the modern theory of the origin of the solar system. That is symmetry, reflection symmetry with respect to the plane of the solar system. Now, if the state of the solar system at any instant is taken as the cause, its state at any later instant is an effect. Thus, the solar system exhibits stability for reflection symmetry. In spite of the various and numerous internal and external perturbations that the solar system has suffered throughout its history, which might be expected to have broken the reflection symmetry more and more as time went on, the symmetry has obviously been preserved. Symmetry in processes will be discussed in detail in Chap. 11.

As an example of lability, consider the DC circuit shown in Fig. 5.4 in Sect. 5.1 as an example of the minimalistic application of the symmetry principle. That system possesses two-fold rotation symmetry (actually, as explained there, functional permutation symmetry). But it is the ideal system that is symmetric. No real circuit built according to the diagram will be exactly symmetric, due to practical limitations on the precision of resistors, of emf sources, and so on. Yet the effect, consisting of the currents and potential differences, always shows two-fold rotation symmetry to reasonable precision. So the situation is certainly not one of instability. Is it stability or lability? To check that, we replace a fixed resistor with a variable one or a fixed emf source with a variable source and gradually break the symmetry of the cause, following the symmetry of the effect with ammeters and voltmeters as we do so. And we find that the symmetry of the effect gets broken as gradually as we break the symmetry of the cause. This proves that the system is labile for two-fold rotation symmetry.

For an example of instability we return to the solar system, this time to its origin and evolution. Modern theory has the solar system originating as a rotating gas cloud with approximate axial symmetry, i.e., symmetry under all rotations about its axis. If that state of what is now the solar system is taken as the cause, the present state can be taken as the effect. Any axial symmetry the protosolar system once possessed has clearly practically disappeared during the course of evolution, leaving the solar system as we now observe it. The random, symmetry breaking fluctuations in the original gas cloud grew in importance as the system evolved, developing into planets and other objects, until the original axial symmetry became hopelessly broken. Additional examples of instability are Birkhoff's 'symmetry paradoxes', referred to previously.

An example exhibiting both stability and instability, although under different conditions, is a volume of liquid at rest in a container. Such a liquid is macroscopically isotropic; its macroscopic physical properties are independent of direction. That is symmetry. The system is stable for isotropy, and the liquid will overcome our attempts to break the symmetry, if only it can. A tap on the side of the container introduces anisotropy, which is very soon damped out by the system, and isotropy is regained. A small crystal of the frozen liquid thrown into the liquid also breaks the symmetry, but it soon melts and isotropy returns. However, when the liquid is cooled to below its freezing point, the situation changes drastically.

Let us imagine that the liquid is cooled very slowly and evenly to below its freezing point and that no isotropy breaking perturbations are allowed; in other words, assume that the liquid is supercooled. It is then still in the liquid state and isotropic. If we now tap the container or throw in a crystal or otherwise introduce anisotropy, the supercooled liquid will immediately crystallize and thus become highly anisotropic. So in the subfreezing temperature range the system's isotropy is unstable; any anisotropic perturbation is immediately amplified until the whole volume becomes anisotropic and stays that way. If, in spite of our precautions, the supercooled liquid should undergo spontaneous symmetry breaking and spontaneously crystallize during or after cooling, the reason is that its own internal random fluctuations are sufficient to set off the instability. The colder the liquid (below its freezing point), the greater the instability. The freezing point is the boundary between the temperature range of stability and that of instability. In fact, it can be defined in that way (see also Sect. 3.2).

The phenomenon of spontaneous magnetization is an additional example of spontaneous symmetry breaking (see Sect. 3.2).

Even in situations of lability and instability, if the perturbation is truly a random one, there exists a sense in which the symmetry of the unperturbed cause, the exact symmetry limit, nevertheless persists as the minimal symmetry of the effect. That pivots on the nature of random perturbations, or, more accurately, on their definition. A random perturbation of an unperturbed cause is defined, for our purpose, as a fluctuating perturbation such that the symmetry of the *totality* of states of the perturbed cause tends to that of the unperturbed cause. In other words, a random perturbation would average out over all states of the perturbed cause to no perturbation at all. One might also express that in terms of observations of the state of the cause, where as the number of observations increases, the symmetry of the *totality* of observed states of the perturbed cause tends to that of the unperturbed cause. Thus, taking the totality of states of a cause as 'the cause' in the symmetry principle, we have that for a random perturbation the totality of states of the effect, as 'the effect' in the symmetry principle, tends to exhibit at least the symmetry of the unperturbed cause. Or, expressed in terms of observation, for a random perturbation, as the number of observations increases, the totality of observed states of the effect will tend to exhibit at least the symmetry of the unperturbed cause.

As an illustration, consider the system consisting of a current of air blowing against the edge of a wedge, with the direction of the air current far upstream from the wedge parallel to the bisector of the wedge, as in Fig. 6.2. Take as the cause the wedge and the incident

Fig. 6.2. Air current blowing against the edge of a wedge, viewed in cross-section

air current a large distance from the wedge, and the air flow around the wedge as the effect. The cause possesses reflection symmetry with respect to the plane bisecting the angle of the wedge. So we expect the air flow above the wedge to be the mirror image of the air flow under the wedge. And, in fact, that is what happens, as long as the speed of the incident air current is sufficiently low so that the flow is laminar. However, at sufficiently high flow speeds vortices form at the edge and are shed downstream. They are not produced symmetrically, but rather alternately at one side and the other. (That periodic phenomenon is the mechanism of production of 'edge tones', the acoustic source of woodwind instruments.)

The effect then apparently possesses less symmetry than the cause. However, random fluctuations in the direction of incident air flow make the reflection symmetry of the cause only an approximate symmetry. At low incident-flow speeds the situation is labile, and the resulting laminar flow around the wedge is also approximately symmetric. (Underblowing a woodwind instrument produces no tone.) At high flow speeds the situation is unstable, and the vortical flow around the wedge lacks reflection symmetry altogether. But if the vortical flow is photographed sufficiently often at random times, the total collection of photographs will exhibit reflection symmetry: For every flow configuration photographed, a mirror image configuration will also appear among the pictures.

The situation, discussed in Sect. 4.5, where the vacuum state of a quantum theory is less symmetric than the rest of the theory, can be viewed as broken symmetry, with the symmetry of the theory without the vacuum state taken as the exact symmetry limit and the vacuum state as the symmetry breaking factor. However, such a situation is sometimes referred to as *spontaneous* symmetry breaking. That nomenclature is misleading, and the attribution of spontaneity is not justified. The vacuum state is an obvious component of the cause for such theories, and its effect in reducing the symmetry of the total cause from that of the exact symmetry limit is in no sense small, nor is it (or should it be) liable to be overlooked.

What might justifiably be called *spontaneous* symmetry breaking in this connection is a higher-level theory, a super theory, that is supposed to explain the first theory, including the lower symmetry of the vacuum state compared with that of the rest of the theory, or at least it is supposed to explain the vacuum state. If the super explanation involves instability and amplification of perturbations, so that the vacuum state, as an effect of the super theory, comes out possessing less symmetry than that of the unperturbed super cause, whatever that might be, then the use of the term 'spontaneous' is reasonable and consistent with our previous discussion. A situation like that is described in Sect. 7.1.

6.3 Summary

In Sect. 6.1 we met with the idea of a metric for a set of states of a system. A metric gives a 'distance' between every pair of states, with null 'distance' for equivalent states and nonzero 'distances' for degrees of inequivalence. Thus, a metric is a softening of the all-or-nothing character of an equivalence relation. That allowed the definition of an approximate symmetry transformation as any transformation that changes every state to a state that is nearly equivalent with the original. An exact symmetry limit for a set of states equipped with a metric and a set of approximate symmetry transformations is another metric for the same set of states for which some of the approximate symmetry transformations (according to the original metric) are exact symmetry transformations. Broken symmetry is another term for approximate symmetry, and a symmetry breaking factor is whatever factor the (possibly hypothetical) switching off of brings about an exact symmetry limit.

We discussed spontaneous symmetry breaking in Sect. 6.2. The crux of the matter was seen to be the issue of stability: How do small, symmetry breaking perturbations of a cause affect the symmetry of its effect? Although, by the symmetry principle, *exact* symmetry of a cause must appear in its effect, *approximate* symmetry of a cause might appear in the effect as exact symmetry (stability), as approximate symmetry (lability), or as badly broken symmetry (instability). The latter situation is what is known as spontaneous symmetry breaking.

Cosmic Considerations

7.1 Symmetry of the Laws of Nature

Let us now return to the analysis of nature into initial state and evolution that we discussed in Sect. 2.5. That reduction, whereby the laws of natural evolution, or laws of nature, and initial states are considered as being independent, seems to be satisfactory for our understanding of nature from the microscopic scale to the macroscopic. It appears to work even for astronomical phenomena, where, for example, we think of different planetary systems or even galaxies as evolving according to the same laws of nature but from different initial conditions. The dichotomy fades away, however, at nature's extreme scales: at the sub-nuclear scale, which is the scale of elementary particles, and at the scale of the Universe. A currently widely accepted cosmological scenario holds that the Universe had its beginning in a 'big bang' in the far past. Can the evolution of the Universe be analyzed in terms of initial conditions and subsequent development according to the laws of nature? Perhaps, if one considers the Universe to be only one of a variety of possible universes, any one of which could have come into being at the time of the big bang. Then the Universe, our universe, would be the result of whatever chance initial state initiated its evolution. And the other potential universes? Are they but theoretical possibilities, or did they all come into being in some sense but are inaccessible to us? Some cosmologists are taking the latter idea seriously [38]. Such speculations, however intriguing, can never be proved or disproved by science, since we cannot experiment with different initial states for the Universe and follow the evolution of each.

Another point of view, which some people find philosophically more satisfactory, is that, since the Universe is by very definition the totality of all we have access to, it is meaningless to consider other possible universes, and we must honor the Universe with the distinction of being the only one possible according to the laws of nature. But what are the laws of nature, if not a description of the evolution of the Universe? What this point of view holds is that the big bang, the Universe, the initial state, and the laws of nature are all intimately intertwined, are all aspects of a single, self-consistent situation. Another aspect of this situation is the variety of elementary particles that nature offers. While cosmologists are asking why the Universe is as it is on the cosmic scale, other scientists are asking why it is as it is on the subnuclear scale, why nature presents us with the kinds of elementary particles it does and no other kinds. Is that phenomenon explainable on the basis of the known laws of nature or must new laws of nature be found that relate more directly to the big bang and its immediately subsequent developments, when the present variety of elementary particles presumably came into existence? The latter possibility seems the more likely. This point of view might be summarized by the statement that things are as they are because they cannot be otherwise [39].

How is symmetry connected with all this? Every aspect of that self-consistent entity, the Universe, exhibits various symmetries and approximate symmetries, and we must try to understand how they, too, fit into the all-encompassing picture. It has even been proposed that the laws of physics are the *result* of symmetries [40, 41]. The following tale is possibly a small part of the picture.

For every allowed state of most of the elementary particles the spatially inverted state (i.e., the state obtained by point inversion, by changing the sign of all three coordinates, x, y, and z, giving what amounts to a rotated mirror image) is also allowed by nature. Now, the law of natural evolution called the strong interaction, which affects certain of the elementary particles, possesses space inversion symmetry (see Sect. 3.1). And all the elementary particles that are affected by the strong interaction are among those that possess space inversion symmetry. Also the law of nature called the electromagnetic interaction is space inversion symmetric. Similarly, all the elementary particles that are affected by this interaction, i.e., all the electrically charged elementary particles, possess space inversion symmetry.

Thus, we have the situation where certain laws of nature and the entities whose behavior they govern possess the same symmetry, space inversion symmetry, also called P symmetry, where P denotes the point inversion transformation.[1] In addition, the transformation of particle–antiparticle conjugation,[2] denoted C, is a symmetry both of the strong and electromagnetic interactions and of the elementary particles whose behavior is governed by those interactions: For every kind of such particle there is a kind that is the antiparticle of it, and all states allowed to particles are also allowed to their antiparticles and vice versa. So again we have the same symmetry, C symmetry, for certain laws of nature and for the objects they apply to.

On the other hand, the neutrinos are deficient in those symmetries. The space inversion image of an allowed neutrino state is not a state allowed by nature. It would be, if the image particle were an antineutrino, for the allowed states of antineutrinos are just the forbidden states of neutrinos. And the inverted states of antineutrinos are forbidden to antineutrinos but allowed to neutrinos. It is as if only left-handed neutrinos and right-handed antineutrinos are permitted (or, perhaps better, required) to exist by nature, while the existence of right-handed neutrinos and left-handed antineutrinos is precluded. Thus, neither P nor C is a symmetry transformation of neutrinos, while the combined transformation of point inversion and particle–antiparticle conjugation, denoted CP, is a symmetry transformation. Now, the strong and electromagnetic interactions do not affect neutrinos, while the law of nature called the weak interaction does. (It also affects other particles.) The weak interaction possesses neither P symmetry nor C symmetry; it does possess CP symmetry.

Are the P and C symmetries of the objects upon which the strong and electromagnetic interactions act basically caused by the P and C symmetries of the interactions themselves? Or vice versa? Or are the interactions and the elementary particles along with their symmetries parts of a self-consistent situation with neither more basic than the other? And are the P and C asymmetries and CP symmetry of the neutrinos fundamentally a result of the P and C asymmetries and CP symmetry of the weak interaction? Or vice versa? Or is it all a self-consistent whole?

---

[1] Actually, P denotes 'parity', which is a property of elementary particles that has to do with what happens to them what a point inversion transformation is applied. But since that is beyond the scope of this book, and since 'point inversion' also starts with P, we can leave the matter as it is.

[2] C actually denotes 'charge' in 'charge conjugation', which is a common term for particle–antiparticle conjugation. Fortunately 'conjugation' also starts with C. So we have reasonable mnemonics.

And how does all that relate to conditions prevailing at the earliest stages of the evolution of the Universe, when the kinds of elementary particles that we know today presumably came into being? Why did the other-handed versions of the neutrinos and antineutrinos not materialize then? Did the laws of nature even then preclude their existence? Or did the chance production of the first neutrino-antineutrino pair as a left-handed neutrino and a right-handed antineutrino cause the as yet 'amorphous' laws of nature to 'crystallize' into such a form that would from that moment onward allow proliferation of the existing versions of neutrino and antineutrino and forbid the existence of the other-handed versions? That would have been a case of cosmic spontaneous symmetry breaking (see Sect. 6.2).

Or should we avoid the question by restructuring our concepts to make CP the only meaningful transformation, so that there would be no reason to consider the hypothetical possibility of other versions of neutrinos and antineutrinos? But then why are P and C separately relevant to the strong and electromagnetic interactions?

Our pursuit of symmetry and of the understanding of symmetry carries us to the very frontiers of modern science and even to speculations out beyond.

7.2 Symmetry of the Universe

Our discussion in Sect. 1.2 led to the conclusion:

Symmetry implies asymmetry.

Or, expressed less succinctly:

Symmetry requires a reference frame, which is necessarily asymmetric. The absence of a reference frame implies identity, hence no possibility of change, and hence the inapplicability of the concept of symmetry.

To see how that works, consider a hypothetically perfectly homogeneous Universe, which might *appear* to possess perfect spatial-displacement symmetry. (In contrast, the real Universe *does* have aspects that are symmetric under spatial displacement.) However, in this version of the Universe there is no reference frame for location, there is nothing to serve as coordinate system. And since the Universe is by definition everything, no externally imposed coordinate system is available. With no coordinate system, all locations are identical. Thus, spatial displacement is meaningless, and so is symmetry or asymmetry under it. This Universe would not possess the possibility of spatial displacement, so the notion of immunity or lack of immunity under such change would be irrelevant to it. We can go even farther and declare that this Universe would have no spatial dimensionality at all! In the absence of anything that could serve as a coordinate system, the claim of identity for all locations is much too mild. More strongly, the very idea of 'location' would be irrelevant to such a universe.

Homogeneity means the possession of identical properties at all locations. Perfect homogeneity means that all locations are absolutely indistinguishable and are thus identical. So there are no different locations at which properties can be compared. All locations are identical and therefore merge into a single location, which makes the very concept of location redundant. And hence no spatial dimensionality. So a perfectly homogeneous Universe is an oxymoron.

How then did we ever come up with the silly idea of a perfectly homogeneous Universe? By extrapolating from the real, inhomogeneous Universe, by trying to imagine the limit of vanishing inhomogeneity. A perfectly homogeneous three-dimensional mathematical space may serve as an approximate model of the real world in certain respects. And we have no difficulty conceptually and meaningfully imposing coordinate systems on it. But as a thing-in-itself, there can be no such animal.

Or, consider nuclear charge symmetry, discussed in Sects. 5.2 and 6.1. Nuclear charge symmetry is explained by assuming that the strong nuclear interaction is blind to the difference between proton and neutron, while the deviation from exact symmetry is explained by the Pauli exclusion principle and the electromagnetic and weak interactions. Note that proton-neutron interchange is possible precisely because there is a difference between proton and neutron. The blindness of the strong nuclear interaction to the proton-neutron difference is an aspect of nature, but so too is the difference between the proton and the neutron an aspect of nature. And it is both the proton-neutron difference and the blindness of the strong nuclear interaction to the difference that bring about the symmetry.

Imagine a hypothetical world in which there is no standard for distinguishing between proton and neutron and thus no difference between them. Then they would be identical, and there would be no protons and no neutrons, only nucleons. That world would not be more symmetric than the real world; it would not possess exact charge symmetry, while in reality our charge symmetry is only approximate. On the contrary, it would have no charge symmetry at all! In that world there would be no possibility of a change to which the strong nuclear interaction would be immune. Replacing a hypothetical nucleon with another hypothetical nucleon would be no change at all. That would be as much a change as replacing a proton with a proton in the real world.

The point here is that a physical change inherently involves a reference frame by which the change acquires significance, a standard to which the change is referred. Indeed, it is the reference frame that makes a change possible. And the reference frame cannot itself be immune to the change under consideration, otherwise it could not serve its purpose.

As we learned in Sect. 1.2, symmetry implies asymmetry, or asymmetry is inherent to symmetry. So if any aspect of the Universe possesses some symmetry, then there must exist another aspect of the Universe that is asymmetric under the change involved in the symmetry. From here follows:

Exact symmetry of the Universe as a whole is an empty concept.

We saw this for perfect spatial displacement symmetry and for perfect charge symmetry in the above examples. Since the Universe is everything, no external reference frames can be imposed on it, and exact symmetry precludes relevant internal reference frames. In the first example, due to its hypothetical perfect homogeneity, i.e., spatial displacement symmetry, the Universe would possess no coordinate system of its own, since a coordinate system would be an inhomogeneity. Thus, there would be no possibility of spatial displacement, so the very concept of spatial displacement symmetry would be inapplicable to the Universe. In the example of nuclear charge symmetry, in a hypothetically perfectly symmetric Universe there would be no standard for differentiating protons from neutrons, as such a standard would introduce asymmetry. Thus, there would be no possibility of proton-neutron interchange, and the very notion of charge symmetry would be irrelevant to such a Universe.

What we have is this:

For the Universe as a whole, undifferentiability of degrees of freedom means their physical identity.

Or, in paraphrase: If it makes no difference to the Universe, then there is nothing else for it to make a difference to.

In the example of a homogeneous Universe, all locations would be undifferentiable and would therefore be identical. As mentioned above, such a Universe would possess no spatial dimensionality at all. We can consider such three-dimensional homogeneous spaces as mathematical models, but only by externally imposing coordinate systems on them. That cannot be done for the Universe. In the charge-symmetric-Universe example, protons and neutrons would be undifferentiable and would thus be identical. We would have only nucleons, which would not be further differentiated.

7.3 No Cosmic Symmetry Breaking or Restoration

The big-bang type cosmological schemes that are currently in vogue generally have the Universe evolve through a number of distinct eras. During each era the Universe evolves in a continuous manner, while the transition from one era to the next is supposed to have the character of a (discontinuous) phase transition, much like crystallization or spontaneous magnetization, which is accompanied by spontaneous symmetry breaking (see Sects. 3.2 and 6.2 and also [42]). One such scheme includes these eras (see, for example, [40] and [43]):

1. Quantum (or Planck) era. The less said the better, to avoid putting one's foot in one's mouth.

2. GUT era. This is the era in which the grand unified theory is assumed to have reigned (see Sect. 5.2). Just as in our era, there were space-time, gravitation, and quantum microscopic behavior. But instead of the three fundamental interactions in addition to gravitation that we now see, there was only a single additional interaction among a single set of elementary matter particles. This interaction was mediated by a single set of force particles.

3. Electroweak era. There were space-time, gravitation, and quantum microscopic behavior. The single nongravitational interaction among elementary particles is assumed to have been replaced by the strong interaction among quarks, mediated by the gluons, and the electroweak interaction among leptons and quarks, mediated by a set of electroweak bosons.

4. Present era. We have space-time, gravitation, and quantum microscopic behavior. We also have the strong interaction among quarks, mediated by gluons, the weak interaction among leptons and quarks, mediated by the W^+, W^-, Z^0 intermediate bosons,

and the electromagnetic interaction among all electrically charged particles, mediated by photons.

During eras 2–4 the interactions are assumed to be describable by gauge theories.

The details and fine structures of the eras and even their number are unimportant for our discussion. So if my list does not fit your favorite scheme, or even if you do not like the names I use, please feel free to make corrections. All we need for our present purpose is a number of temporally ordered eras of continuous evolution preceded by a practically unmentionable era, where each era is the result of a (discontinuous) phase transition from the preceding era.

Now, what do we mean by phase transition? Two often used examples are crystallization and spontaneous magnetization, as I mentioned above and which we discussed in Sects. 3.2 and 6.2. As a material in liquid state is cooled, its properties change continuously until (under suitable conditions) it spontaneously and discontinuously crystallizes into a solid state. Or, as an unmagnetized ferromagnet is cooled, various of its properties change continuously until it spontaneously and discontinuously becomes magnetized.

In each of the examples the phase transition is, and phase transitions in other systems are often, accompanied by spontaneous symmetry breaking, whereby equivalent degrees of freedom suddenly become inequivalent (and the set of symmetry transformations is reduced). In crystallization the system jumps from a state of no distinguished positions and directions to a state of distinguished positions and directions. The choice of those distinguished positions and directions is extremely sensitive to conditions and is thus effectively random. The symmetry of an effectively homogeneous and isotropic medium is broken to that of a crystal lattice. In magnetization the system jumps from a state of no distinguished direction to one of a single distinguished direction, the direction of magnetization, whose choice is again extremely sensitive to conditions and is thus effectively random. The symmetry of an effectively isotropic medium is broken to that of a vector, or a directed line. (There is a complication here due to the character of a magnet as equivalent with a current loop, but we can ignore that now.)

In each example and in general, at a symmetry breaking phase transition the volume of the system might divide into domains, whereby the symmetry breaking takes different directions in different domains. In crystallization the resulting solid might be composed of crystalline domains, of which the crystal axes are differently oriented in each.

And in spontaneous magnetization the ferromagnetic medium might divide up into magnetic domains, where the direction of magnetization is different in different domains. Adjacent domains are separated by relatively thin transition surfaces called domain walls.

That is what can happen at ordinary-scale phase transitions during the present cosmological era. But what happened at the assumed *cosmic* phase transitions, when the *whole Universe* is assumed to have jumped from its state at the end of one era to its state at the beginning of the next? Consider, for example, the transition from the GUT era (2) to the electroweak era (3). The major change seems to have been that the single interaction among a single set of matter particles, mediated by a single set of force particles, became the strong interaction among quarks, mediated by gluons, along with the electroweak interaction among leptons and quarks, mediated by a set of electroweak bosons.

Does that mean the quarks evolved from an equal number of prequarks, the leptons evolved from an equal number of preleptons, and the prequarks and preleptons were somehow equivalent in era (2) but became inequivalent at the beginning of era (3)? Does that mean the gluons evolved from an equal number of pregluons, the electroweak bosons evolved from an equal number of pre-electroweak bosons, and the pregluons and pre-electroweak bosons were somehow equivalent in era (2) and became inequivalent at the beginning of era (3)? Does that mean the family of symmetry transformations of era (3) was reduced from the family of symmetry transformations of the GUT gauge theory?

No! Our discussion in Sect. 7.2 taught us that cosmic equivalence means identity. In era (2) there simply was no reference frame by which prequark-prelepton and pregluon-pre-electroweak-boson distinctions could have been possible, so there was no distinction. And that means identity, not equivalence. Thus, there were no equivalent prequarks and preleptons that at the phase transition became inequivalent quarks and leptons. There was only a set of GUT matter particles, which at the phase transition was replaced by a set of quarks and leptons. And similarly, there was only a set of GUT bosons, which at the phase transition was replaced by a set of gluons and a set of electroweak bosons. The number of members of each GUT set was not simply the sum of the numbers of members of the two respectively resulting sets of era (3). The particle menagerie of era (2) is open to speculation, although I would assume that the numbers of particle kinds were likely to be less than the just-mentioned sums. (This

assumption is based on my personal gut feeling.) Indeed, the particle
zoo of the electroweak era (3) is similarly open to speculation, and I
would similarly assume that the numbers of particle kinds then were
likely less than their numbers in the present era (4).

The questions posed in the second preceding paragraph are the
result of extrapolating backward in time from the reference frame of
era (3). Such extrapolation gives meaning to the terms used in those
questions. However, such conceptual extrapolation on our part in no
way obliges era (2) to conform. The transition from era (2) to era (3)
is supposed to have been discontinuous, so a continuous conceptual
limiting process from era (3) back to era (2) is in principle useless.

In what sense, then, can the transitions from era to era be viewed as
phase transitions? Clearly not in the sense of the equivalent becoming
inequivalent, i.e., not in the sense of symmetry breaking. Symmetry
change, yes. Each era had and has its own characteristic symmetry
expressed in terms of the degrees of freedom of that era. But the change
in symmetry at a transition was not symmetry breaking; it was not
a reduction of the group of symmetry transformations. At least not if
we want to retain the discontinuity of the transitions. If we choose to
give up discontinuity, the cosmological scheme will have an altogether
different character, and the concept of phase transition will not be
applicable. If, however, we keep discontinuity, then the only sense in
which the transitions might be considered phase transitions is in their
discontinuous character itself. Yet, as a compensating factor that we
saw above, we have the appearance of *new degrees of freedom*.

A by-product of discontinuous transition and the appearance of new
degrees of freedom is, in analogy to what happens in laboratory phase
transitions, the possibility of domaining. Space might become divided
into domains, in which the 'orientation' of physics in the abstract space
of the new degrees of freedom is different in different domains. Such
domains would be separated by relatively thin domain walls, which
might be of importance for the formation of galaxy clusters. It is hard
to see how thin domain walls could form as a result of continuous
cosmic evolution.

Physics involves the devising of metaphors to describe reality. Our
metaphors are often mathematical, but still they are metaphors. 'Phase
transition' is a metaphor for describing the transitions between the eras
of big-bang type cosmological schemes. In order not to be misled, we
are well warned not to take this metaphor (or any metaphor, for that
matter) too literally. As we just saw, the 'phase transition' metaphor is

appropriate only in that discontinuity and the possibility of domaining are common to both cosmic and ordinary-scale phase transitions. It is inappropriate in that, while ordinary-scale phase transitions might involve the equivalent becoming inequivalent, i.e., they might involve symmetry breaking, cosmic transitions cannot involve the equivalent becoming inequivalent and thus cannot involve symmetry breaking.

The reasoning of this and the preceding section leads to the following interrelated conclusions:

1. *Cosmological schemes cannot involve perfect symmetry for the Universe as a whole.*

Thus, no symmetry we consider for the present cosmic era (4), be it the SU(3) gauge symmetry of the strong interaction or any other, can be assumed to be perfect. Some aspect of the Universe must violate it. And the same for the previous cosmic eras.

2. *Cosmological schemes cannot involve fundamentally undifferentiable, yet still somehow different, degrees of freedom of the Universe.*

We might try to imagine such degrees of freedom for previous cosmic eras by conceptually imposing upon those eras the reference frame of the present era. But that is physically meaningless, since the reference frame of the present era was not part of the Universe then. For an additional example, it is assumed that during era (3) the present electromagnetic and weak interactions were unified as a single interaction, the electroweak interaction. Then, it is assumed, the precursors of the Z^0 intermediate boson of the weak interaction and the photon, the mediator of the electromagnetic interaction, as different as the latter two are in the present era (4) (e.g., the Z^0 possesses mass, while the photon is massless), were somehow undifferentiable while still forming two degrees of freedom. That is meaningless.

3. *Cosmological schemes with phase transitions between eras cannot involve symmetry breaking.*

If a transition was continuous, then perfect symmetry could not have become approximate symmetry. And according to our conclusion, there could not have been perfect symmetry anyway. However, approximate symmetry could have changed its approximation at a continuous cosmic transition. Thus, at a continuous cosmic transition a good approximation could have worsened, perhaps *imitating* symmetry breaking.

If a transition was discontinuous, was a 'phase transition', the character and number of degrees of freedom could have changed. Thus, one (approximate) symmetry could have changed to another. But undifferentiable degrees of freedom becoming differentiable could not have occurred, since there could not have been undifferentiable degrees of freedom to begin with. So no symmetry breaking. At a discontinuous cosmic transition, at a 'cosmic phase transition', there could have occurred symmetry change, but no symmetry breaking.

It then follows that cosmological schemes that assume perfect symmetry of, or equivalently, indistinguishable degrees of freedom for, the Universe are meaningless. I am not claiming that such schemes cannot be perfectly valid schemes by the criteria of consistency with experimental data and self-consistency. Neither am I claiming that such schemes cannot be very useful and valuable in addition to their being beautiful and amazing intellectual achievements. Nevertheless, when cosmological schemes assume perfect symmetry of the Universe, they are indeed meaningless.

One way of circumventing the meaninglessness is to take such schemes as *approximate* descriptions of a situation that is not perfectly symmetric, just as a spatially homogeneous model is taken as an approximation to describe the real Universe. A price to pay for that is giving up the idea, if one in fact holds the idea, that such schemes could be final and exact descriptions of the Universe.

It is commonly taken for granted that by raising particle accelerator energies higher and ever higher, thus probing physics at higher temperatures, at shorter distances, and at shorter time intervals, we are actually investigating the conditions prevailing during previous cosmic eras. Indeed, it is assumed that if we managed to produce energies high enough to probe time intervals and distances at the Planck scale (about 10^{-43} second and 10^{-35} meter), we would even be investigating the quantum era (1) itself. However, the idea that we can reconstruct past cosmological eras by investigations performed in the present era is a fallacy, as long as we are assuming discontinuous cosmic transitions.

The problem can be expressed in this way: Why should the high-energy physics of the present era reflect the physics of previous eras? For a model of continuous cosmic evolution that would indeed be a reasonable assumption; we might then very well assume that by raising accelerator energies we would be reconstructing previous cosmic conditions in our laboratories. But discontinuous transitions are barriers to such 'time travel'. The reconstruction idea is reasonable only as far back as the beginning of the present era. Beyond that it just does not hold water.

The essence of the matter is that the physics of the present era, however high the energy might be, is still a characteristic of the present era. It is occurring in the context of the reference frame of the present era. For example, no matter how similar the photon and the Z^0 become at higher and higher energies, they never become identical and are always distinguishable in principle. However, in the electroweak era (3), the era immediately preceding the present era, the situation was *qualitatively* different from that of the present era, as we have learned above. 'Indistinguishable' as the limit of 'barely distinguishable' is very different from 'identical'.

The assumption of reconstruction, or symmetry restoration, is carrying the 'phase transition' metaphor too far. It is true that by reheating a crystalline solid or magnetized ferromagnet we restore the symmetry that was broken at the phase transition induced by cooling. The cosmic analog would be the reheating of the whole Universe. And that is an extremely far cry from the high-energy physics of the present era, in which an infinitesimal part of the whole Universe, merely a few or even a few hundred particles within an infinitesimal volume of space, are heated infinitesimally briefly within a cold environment.

It is not unreasonable to expect that as we go to higher energies, new symmetries will turn up, so that higher-energy physics will be characterized by a higher degree of symmetry than lower-energy physics is (and the family of symmetry transformations of the latter will be reduced from that of the former). It is not unreasonable to hope that at sufficiently high accelerator energies the weak and electromagnetic interactions will be found to merge into a unified electroweak interaction, whose symmetry subsumes that of the distinct interactions. But it is completely baseless to assume that we are thus reconstructing past eras and thus restoring the symmetries that were assumed broken at the cosmic phase transitions. (In fact, as we saw above, there can be no symmetry breaking at discontinuous cosmic transitions.) Specifically, there is no reason whatsoever to expect that the electroweak interaction and its symmetry that we might discover at sufficiently high energies should reflect the actual situation during era (3).

Yet, in order to construct *some* cosmological scheme rather than simply giving up in despair, we might, not unreasonably, assume that high-energy physics does give us some indication, however imperfect, of the situations in previous cosmic eras. We know we cannot count degrees of freedom. But perhaps we can deduce the general character of the situation. Indeed, that is how the eras presented at the beginning of this section were proposed. For example, it is not unreasonable to assume that the era preceding the present one was characterized by, among its other characteristics, an interaction additional to and weaker than the strong interaction and a set of mediating bosons (distinct from gluons) that affected via that interaction both quarks and another set of matter particles that were lighter than quarks. That interaction is assumed to have 'split' into the weak and electromagnetic interactions at the cosmic phase transition between the preceding era (3) and the present one (4), and so it is reasonable to call the interaction 'electroweak' and the additional set of matter particles 'leptons'. However, that interaction is *not* the expected high-energy merger of the electromagnetic and weak interactions. It is altogether another animal. That interaction is assumed to have been a characteristic of the preceding cosmic era, while the latter is expected to be a property of the present era.

High-energy physics cannot be expected to reflect precisely the situation that prevailed during earlier cosmic eras that evolved into the present era via phase transitions, although it might be indicative. Specifically, any symmetry emerging at high energies cannot have been a feature of such earlier eras.

I would like to emphasize, however, that if the evolution of the Universe occurred in a continuous manner, instead of via (discontinuous) phase transitions, then an approximate symmetry for the Universe in some era could have worsened in a later era, and high-energy physics in our era might indeed reveal the situations that prevailed in earlier cosmic eras. In either case, whether cosmic evolution proceeded discontinuously or continuously, or whether what actually took place is better described in different terms altogether, the results of future high-energy physics experiments, such as at CERN's Large Hadron Collider, are eagerly awaited by physicists, at least by those specializing in such matters. The experimental results might give physicists a better understanding of the Universe in our own era. And they might shed light on earlier eras as well. However, they might instead throw physics into turmoil by posing more puzzles than they solve. We wait and see.

7.4 The Quantum Era and The Beginning

As I mentioned in Sect. 7.3, it is commonly assumed that if we succeeded in probing the Planck scale, we would be investigating the quantum era (1). Nevertheless, as we saw in Sect. 7.3, that assumption is fallacious. Yet, can anything, however general and qualitative, be reasonably deduced concerning the quantum era? What can reasonably be thought to have preceded the GUT era (2), assuming, of course, that there was indeed a GUT era and that it was the result of a discontinuous cosmic transition?

So let us assume there was a GUT era characterized by space-time, gravitation, quantum microscopic behavior, and, say (the details are not important), a single interaction, mediated by a single set of force particles, among a single set of matter particles. And let us consider what the high-energy physics of the present era tells us. What? We have not reached the Planck scale yet? What shirkers those experimentalists are! Never mind. Let us consider what we *think* we would find at the Planck scale. We think that at the Planck scale we would discover the fundamental quantum character of space-time, also called quantum gravity. We expect to find quantum fluctuations of space-time itself, a situation suggestively called 'space-time foam' [44]. We expect to find some of the fluctuations leading to the 'pinching off' of Planck-size regions, which become disconnected from the Universe and form 'baby universes' [38]. What those picturesque, vaguely meaningful metaphors indicate is that we think known physics, including the notion of space-time itself, utterly breaks down at the Planck scale.

Now, the assumed transitions from era (2) to era (3) and from era (3) to era (4) had the property of carrying a situation that can be considered simpler into one we might deem more complex. A single interaction in era (2) became two interactions in era (3), which then became three in the present era (4). Using that as a guide, we expect the quantum era (1) was somehow simpler than the GUT era (2). In what way simpler? One interaction less than a single interaction is no interaction. Perhaps some protogravitation in era (1) can be viewed as splitting into gravitation and the single interaction of era (2). But gravitation is intimately connected with space-time. And the assumed results of our gedanken Planck scale high-energy investigations point to the irrelevance of space-time, as we are macroscopically familiar with space and time, to the quantum era. So then macroscopic gravitation, or something like it, appears to be out as well.

It looks as if our surest guesses about the character of the quantum era are negative: no space, no time, no gravitation. How, then, can we conceive of *anything* about the quantum era, if we cannot do so in terms of space and time, in terms of being and becoming, i.e., in terms of existence and change? (Our metaphoric description of the Planck scale breakdown of known physics was couched in terms of space and time, of being and becoming.) One might try something like this: "The quantum era was a situation of highly quantum character, strongly fluctuating. It was unstable to fluctuations and thus underwent a transition to era (2) and space-time." But the idea of instability leading to transition implies becoming and time.

In the cosmological scheme of eras (1)–(4) certain properties of certain eras are supposed to have carried over, fully or partially, into the subsequent eras. For example, space-time is assumed to have carried over from era (2) to era (3) and on to the present era (4). And something of the electroweak interaction of era (3) is supposed to be reflected in the present electromagnetic and weak interactions, especially in their high-energy behavior. Furthermore, something of the assumed grand unified interaction of era (2) is supposed to be reflected in the present strong, weak, and electromagnetic interactions. The assumed describability of the era (2) interaction by a gauge theory formulated in spatio-temporal terms seems to have carried over fully into the present era, since all three present interactions appear to possess that character. And presumably the very-high-energy physics of the present era should reflect other relic properties of the GUT era interaction.

Now, the quantum era, too, presumably bequeathed properties to its descendants. The moderate quantum character of the present era – moderate, because it is not dominant at all scales but mostly only at the submicroscopic scale – might be thought of as a relic of an extreme quantum character of era (1). And the assumed nonspatiality and nontemporality of the quantum era might be considered to be the source of present quantum spatial and temporal nonlocality. The idea here is that according to quantum theory all locations and all times, separately, are in a certain sense equivalent. In the quantum sense all locations can be thought of as *the same location* and all times as *the same time*. Thus, for example, the fact that a measurement at one place instantaneously 'affects' the situation at other places can be understood, rather than as faster-than-light propagation, better as *no propagation at all*. The 'effect' of the measurement does not have to go anywhere; it is already there, since there and here are in a sense the same. But if all locations are the same location and all times the same time, the situation is reduced, in the relevant quantum sense, to zero spatiotemporal dimensionality. In other words, to nonspatiality and nontemporality.

On the other hand, as we saw above, it is reasonable not to consider space-time to be a property of the quantum era, so the spatiotemporal character of subsequent eras cannot be thought of as a quantum-era relic. The origin of space-time should then be understood to be the spontaneous appearance of new degrees of freedom at the transition from the quantum era to era (2). And those degrees of freedom are assumed to have survived the transitions from era (2) to the present.

What else can be said about the quantum era? Very little of any physical significance, it seems to me. I have emphasized elsewhere [45] that cosmological schemes, dealing as they do with a unique phenomenon par excellence, the Universe as a whole, have exceeded the domain of physics and have ventured into the domain of metaphysics. That is true a fortiori for considerations involving the quantum era. I do not intend to imply that cosmological schemes do not involve physics nor that they are not very useful for physics. Indeed, a successful cosmological scheme would be a marvelous achievement and would offer physicists important and useful insight and guidance. Yet, given the inaccessibility of the quantum era from the present era and the current status of our cosmological understanding, it seems reasonable that the more detailed any statement about the quantum era is, the more suspect that statement should be held to be. For example, specific equations have been proposed to describe the quantum era. Such considerations actually belong to the domain of metaphysics, and, although expressed in the language of physics, they really have little, if any real physics content.

That brings us to the subject of The Beginning. It seems to be a common misconception that cosmological schemes of the general type of that presented in Sect. 7.3 imply this chronological sequence: (a) The Beginning, followed by (b) the quantum era, which had a duration of about 10^{-43} second, which in turn was followed by (c) era (2), and so on. However, as we saw above, the quantum era seems best considered nontemporal. Thus, it should not be thought of as having been characterized by any time duration at all. The Planck time of about 10^{-43} second is considered to be characteristic of the quantum nature of space-time in the present era. But the quantum era is not the present era, nor is it reasonably considered to have possessed the property of time. The assignment of duration to the quantum era is an unwarranted extrapolation from the present era to the quantum era.

It is a conceptual imposition of a reference frame of our era on an era inherently possessing no such reference frame.

It also follows from the nontemporality of the quantum era that it cannot be thought of as having been preceded nor as having been succeeded by anything. From its own, nontemporal point of view the very concepts of precession and succession are meaningless for the quantum era. However, the quantum era can still be considered to have been followed by era (2) in the following carefully construed sense. Era (2) is assumed to have been characterized by time. Thus, *from the temporal reference frame of era (2)* the quantum era can legitimately be thought of as having preceded era (2), just as era (3) is thought of as having followed era (2). Then by verbal manipulation we replace the expression 'the quantum era preceded era (2)' with the expression 'era (2) followed the quantum era'. But in both cases the temporal ordering is with respect to the reference frame of era (2).

Thus, the quantum era, by its reasonably assumed nontemporality, forms a barrier to the flight of our imagination back in time in search of The Beginning. Although it can be thought of as having preceded era (2), it itself cannot be considered as having had duration. Nor are the concepts of 'the beginning of the quantum era' or 'before the quantum era' anything but vacuous. So The Beginning, as the beginning of the quantum era or as whatever preceded the quantum era, is utterly meaningless.

A reasonable alternative to The Beginning is 'the beginning of time', in whatever sense the latter can be assigned meaning. Now, since the quantum era can be thought of as having preceded era (2), and since era (2) is thought of as having been characterized by time, the quantum era itself or the transition from the quantum era to era (2), the transition to space-time, might be thought of as the beginning of time. The idea is that the quantum era and the transition to space-time are considered to precede any time. As far back in time as we imagine – and using a suitable time variable, we can imagine going back in time 'forever' [46] – the quantum era and the transition to space-time will still be considered to be earlier. That is the meaning we can assign to 'the beginning of time'. So if one feels any need for The Beginning, the quantum era or perhaps the transition to space-time can reasonably fulfill that need. In summary:

> *The quantum era can reasonably be assumed to have been nontemporal, nonspatial, and extremely quantal. The Beginning can*

reasonably be identified with the quantum era or with the transition to space-time.

Let me remind the reader that the concept of a quantum era is a component of big-bang type cosmological schemes, which are attempts to model and grasp the evolution of the Universe. At present, such schemes seem to be the best we have. However, their validity is not guaranteed. Whether the evolution of the Universe really occurred in such a way is still an open question, and the existence of a quantum era is by no means assured. More experiments, additional observations, and further theoretical research will surely improve our cosmic understanding, and we look forward to future developments.

But to put matters in perspective, think how humanity's cosmology has evolved over the millennia, over the recent centuries, and even during the past century. At every stage the Universe was understood in terms and in a manner that were appropriate and valid for that era [47]. When the cosmological picture eventually changed over time, people would deride the primitive concepts of their ancestors, just as we look back even as recently as to the early twentieth century and wonder at the limited thinking of cosmologists then. Considering that science in general and cosmology in particular are accelerating in pace, I strongly suspect that in only some ten to twenty years, *we* will be the ones who will serve as objects of ridicule: What idiots; how could they have thought *that*? Well, we will see in some ten to twenty years.

7.5 Summary

In Sect. 7.1 we looked into the significance of reduction to initial state and evolution for the Universe as a whole. Since we have access to only a single universe, that reduction is problematic. We also considered how symmetry of the laws of nature might fit into the picture.

We discussed symmetry of the Universe in Sect. 7.2 and reached the conclusions that the Universe as a whole cannot possess exact symmetry, and for the Universe as a whole, undifferentiability of degrees of freedom means their physical identity.

In Sect. 7.3 we considered big-bang cosmological schemes that have the Universe evolve through a number of eras, starting with a 'quantum era'. The discussion led to these conclusions: (1) Cosmological schemes cannot involve perfect symmetry for the Universe as a whole. (2) Cosmological schemes cannot involve fundamentally undifferentiable, yet still somehow different, degrees of freedom of the Universe. (3) Cosmological schemes with phase transitions between eras cannot involve symmetry breaking. (4) High-energy physics cannot be expected to reflect precisely the situation that prevailed during earlier cosmic eras that evolved into the present era via phase transitions, although it might be indicative. Specifically, any symmetry emerging at high energies cannot have been a feature of such earlier eras. However, if the evolution of the Universe occurred in a continuous manner, instead of via (discontinuous) phase transitions, then an approximate symmetry for the Universe in some era could have worsened in a later era, and high-energy physics in our era might indeed reveal the situations that prevailed in earlier cosmic eras.

Finally, we speculated in Sect. 7.4 on the nature of the quantum era that presumably served as the first in the sequence of eras forming the evolution of the Universe and considered what The Beginning of that evolution could have been.

The Mathematics of Symmetry: Group Theory

So far in this book I have been trying to deal with symmetry conceptually and make do with as little formalism and mathematics as reasonably possible. I must admit that, especially in our discussion of the symmetry principle in Chap. 4, I needed to make use of heavier formalism than I was happy doing. But I thought the subject was too important to postpone to after the formalism chapters.

Nevertheless, when it comes to seriously applying symmetry considerations in science, the conceptual approach can go only so far, which is not very far. Particularly when the applications are quantitative, the conceptual approach is simply incapable of supplying the necessary tools. So for the application of symmetry in science it is necessary to develop a general symmetry formalism. In this book the formalism is developed in Chap. 10. It is couched in the mathematical language of symmetry, which is group theory. So in order to get a grasp of the symmetry formalism, it is necessary to obtain some grasp of group theory. That is what we will aim for in this chapter and the next.

The present chapter presents the barest minimum of the most fundamental ideas of group theory. In Chap. 9, we will build on that foundation to achieve what I think is a reasonable introductory body of understanding.

8.1 Group

A *group* is, first of all, a *set* of *elements*. In a more rigorous presentation one might first study some set theory before approaching group theory. However, for our purposes it is sufficient to know that a set is a collection and the elements of a set are what are collected to form the collection. An element of a set is said to be a member of the set, or to belong to the set. Besides whatever property the elements of a set possess that makes them belong to the set, they need have no additional properties. For a set to be a group, however, very definite additional properties are needed, as we will see in the following discussion.

The number of elements of a set or group is called its *order*. It may be finite, denumerably infinite, or nondenumerably infinite. You know what finite means, and you most likely have a reasonable idea of what infinite means. *Denumerably* infinite means that the elements can be labeled by the natural numbers, 1, 2, 3, ... So the set of natural numbers is by definition denumerably infinite, and so is the set of all integers and even – believe it or not – the set of all rational numbers (numbers expressible as the ratio of two integers), as examples. *Nondenumerably* infinite means that the elements cannot be labeled by the natural numbers. The set of all real numbers, for instance, is of nondenumerably infinite order, as are the set of all straight lines in a plane and the set of all spheres in three-dimensional space.

We use italic capital letters to denote sets and groups and italic small letters for elements, in general. Curly brackets { } denote the set consisting of all elements indicated or defined within them. Thus, the equation

$$S = \{a, b, c, d\}$$

means that the set denoted by S consists of the elements denoted by a, b, c, d. The sign $=$ in such equations means that the set on the left-hand side and that on the right are one and the same. Another example is

$$P = \{\text{all real numbers } a \text{ such that } -1 < a < +1\},$$

which means that the set P consists of all real numbers greater than -1 but less than $+1$.

The symbol $\in$ is used to indicate membership in a set or group. The statement

$$w \in U$$

means that the element w is a member of the set U.

Another use of the sign $=$ is to relate elements of a set, as in the equation

$$a = b.$$

Here the $=$ sign means that the two elements a and b are one and the same element.

For a set to be a group it must be endowed with a *law of composition*. What this means is that any two elements of the set can be combined (without specifying just how that is to be done). In fact, any pair of elements can be combined in two ways. If a, b are elements of G, the two compositions of a and b according to the law of composition of set G are denoted ab and ba. (It is fortunate with regard to notation that composition involves only two ways of combining!)

But in itself composition does not yet make a group. To be a group, a set G together with its law of composition must satisfy the following four properties:

1. *Closure.* For all a, b such that $a, b \in G$, we have

$$ab, \, ba \in G \, .$$

   This means that for all pairs of elements of G both compositions are themselves elements of G. Another way of stating this is that the set G is closed under composition.

2. *Associativity.* For all a, b, c such that $a, b, c \in G$, we have

$$a(bc) = (ab)c \, .$$

   This means that in the composition of any three elements the order of combining pairs is immaterial. Thus, one can evaluate abc by first making the composition $(bc) = d$ and then forming ad, corresponding to $a(bc)$, or one can start with $ab = f$ and then make the composition fc, corresponding to $(ab)c$. Both results must be the same element of G, which can thus unambiguously be denoted abc. In short, one says that the composition is associative. It then follows that associativity holds for composition of any number of elements.

3. *Existence of identity.* G contains an element e, called an *identity* element, such that

$$ae = ea = a \, ,$$

   for every element a of G. The characteristic property of an identity element is, then, that its composition in either way with any element of G is just that element itself.

4. *Existence of inverses.* For every element a of G there is an element of G, denoted a^{-1} and called an *inverse* of a, such that

$$aa^{-1} = a^{-1}a = e \,.$$

In words, for every element of G there is an element whose composition with it in either way is an identity element. (It might happen that $a^{-1} = a$ for some or all elements a of a group.)

That ends the definition of a group. To summarize, a group is a set endowed with a law of composition such that the properties of (1) closure, (2) associativity, (3) existence of identity, and (4) existence of inverses hold. A group is abstract if its elements are abstract, i.e., if we do not define them in any concrete way. However, with an eye to the future, we will take special interest in groups whose elements are transformations. For such groups the law of composition of two transformations is consecutive application of the transformations. That is discussed in detail in Sect. 10.2.

In general $ab \neq ba$ in a group. (Otherwise, why insist on two ways of composition?) But it might happen that certain pairs of elements a, b of G do obey $ab = ba$. Such a pair of elements is said to *commute*. From property 3 of the definition of a group it is seen that an identity element e commutes with all elements of a group and that every element commutes with its inverse. Obviously, every element of a group commutes with itself. If all the elements of a group commute with each other, i.e., if $ab = ba$ for all elements a, b of G, the group G is called *commutative* or *Abelian* (after the Norwegian mathematician Niels Henrik Abel, 1802–1829, but do not conclude that everyone who studies group theory dies so young).

In property 3 we demand the existence of an identity element but do not demand that it be unique. And in property 4 and the paragraph preceding the present one I carefully refrain from referring to *the* identity in order not to imply uniqueness. However, all that deviousness was only to allow us the pleasure of *proving* that the identity is indeed unique. This will be our first example of a group theoretical proof. Please note that we will be very careful to justify each operation and each equation by reference to the definition of a group. That is especially important, since we are using familiar notation (multiplication, parentheses, equality) but assigning it novel significance that takes getting used to.

To prove uniqueness of the identity, we assume the opposite, that more than a single identity element exist, and show that this leads to a contradiction. Denote any two of the assumed different identity elements by e' and e''. By property 3 we then have

$$e'a = ae' = e''a = ae'' = a \, ,$$

for every element a of group G. Evaluate the relation $e'a = a$ for the specific element $a = e''$. We obtain $e'e'' = e''$. Now evaluate $ae'' = a$ for $a = e'$. We get $e'e'' = e'$. Comparing the two results, we have $e' = e''$, which is a contradiction to our assumption that e' and e'' are different. That proves the uniqueness of the identity.

Although we do not require it in property 4, the inverse of an element is unique; i.e., for every element a of group G there is only one element, denoted a^{-1}, such that

$$aa^{-1} = a^{-1}a = e \, .$$

As an additional example of group theoretical proof, we prove that statement. We again assume the opposite and show how it leads to a contradiction. So assume that element a possesses more than one inverse and denote any two of them by b, c. Then by property 4

$$ab = ba = ac = ca = e \, .$$

Now, according to property 2, the associativity property,

$$c(ab) = (ca)b \, .$$

By the equation before last and using the property of the identity, the left-hand side of the last equation is

$$c(ab) = ce = c \, ,$$

while the right-hand side becomes

$$(ca)b = eb = b \, .$$

So the associativity property gives $b = c$, and we have a contradiction to our assumption that b and c are different. That proves uniqueness of inverses.

From the definition and uniqueness of inverses it is clear that the inverse of the inverse is the original element itself, i.e., the inverse of a^{-1} is a. In symbols,

$$(a^{-1})^{-1} = a \, .$$

The inverse of a composition of elements is the composition in opposite order of the inverses of the individual elements. In symbols this means that

$$(ab)^{-1} = b^{-1}a^{-1} ,$$

$$(abc)^{-1} = c^{-1}b^{-1}a^{-1} ,$$

and so on, which is verified directly. For example, we show that $b^{-1}a^{-1}$ is indeed the inverse of ab by forming their composition and proving it is the identity:

$$(ab)(b^{-1}a^{-1}) = a(bb^{-1})a^{-1} \quad \text{(by associativity)}$$

$$= aea^{-1} \quad \text{(by inverse)}$$

$$= (ae)a^{-1} \quad \text{(by associativity)}$$

$$= aa^{-1} \quad \text{(by identity)}$$

$$= e \quad \text{(by inverse)} .$$

The verification of

$$(b^{-1}a^{-1})(ab) = e$$

is performed similarly.

We define powers of elements by

$$a^2 = aa , \qquad a^3 = a^2a = aa^2 = aaa ,$$

and so on, and

$$a^{-2} = a^{-1}a^{-1} = (a^2)^{-1} ,$$

and so on. Then the usual rules for exponents are largely applicable, except that noncommutativity must be kept in mind. For example, if elements a, b do not commute with each other, then

$$(ab)^2 = (ab)(ab) = abab \neq a^2b^2 .$$

The *structure* of a group is a statement of the results of all possible compositions of pairs of its elements. For finite-order groups that is most clearly done by setting up a *group table*, similar to an ordinary multiplication table. Refer to Fig. 8.1. To find ab look up a in the left column and b in the top row; the composition ab is then found at the intersection of the row starting with a and the column headed by b.

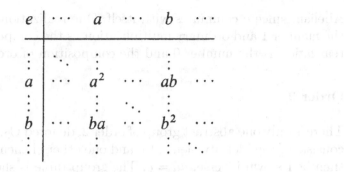

Fig. 8.1. Group table

For the composition ba it is b that is on the left and a at the top. When the symbols for the group elements have the same ordering in the top row (left to right) as in the left column (up to down), a group table will be symmetric under reflection through the diagonal if and only if the group is Abelian. That is easily seen from the figure.

Note that by reordering the rows and columns or by changing the symbols denoting the group elements, a group table can be made to look different while still describing the same group. That must be taken into account when comparing group tables. Thus, two group tables describe the same abstract group, i.e., express the same structure, if they can be made identical by reordering rows and columns and redenoting elements.

For groups of infinite order a group table is obviously impossible. Instead, the results of all possible compositions must be expressed by a general rule.

A concrete example of an abstract group, i.e., a group of concrete elements with a concretely defined law of composition, possessing the same structure as an abstract group, is called a *realization* of the abstract group. Such realizations might be, for example, groups of numbers, of matrices, of rotations, or groups of other geometric transformations.

We now consider all abstract groups of orders 1 to 5 and one abstract group of order 6, and present examples, or realizations, of each but the last.

Order 1

There is only a single abstract group of order 1. Its symbol is C_1. It is the trivial group consisting of only the identity element e. It is

Abelian, since e commutes with itself. One realization of the group is the number 1 and ordinary multiplication as the composition. Another realization is the number 0 and the composition of ordinary addition.

Order 2

There is only one abstract group of order 2, denoted C_2. It is Abelian. It consists of the identity element e and one other element a, which must then be its own inverse, $aa = e$. The group table is shown in Fig. 8.2. A realization of the group is the set of numbers $\{1, -1\}$ (the two square roots of 1) under ordinary multiplication. The number 1 serves as the identity, while the number -1 is its own inverse, $(-1) \times (-1) = 1$. Another, geometric realization is the set consisting of the transformation of not doing anything, called the *identity transformation*, and the transformation of mirror reflection (in a two-sided mirror), the composition being consecutive reflection. (Transformations are discussed in more detail in Sect. 10.2.) The identity transformation serves as the group's identity element, and the reflection transformation is its own inverse, since two consecutive reflections in the same mirror bring the situation back to what it was originally.

Another realization is the set of rotations about a common axis by $0°$, which is another way of expressing the identity transformation, and by $180°$, with the composition of consecutive rotation. Note that the group elements here are not the orientations of $0°$ and $180°$, but the rotations through those angles. Rotation by $0°$ is the identity element. Rotation by $180°$ is its own inverse, since two consecutive rotations by $180°$ about the same axis amount to a total rotation by $360°$, which is no rotation at all, the identity element.

Fig. 8.2. Group table of C_2

Order 3

There is only a single abstract group of this order, denoted C_3, and it is Abelian. It consists of the identity e and two more elements a and b, each of which is the inverse of the other,

$$ab = ba = e ,$$

and each of which composed with itself gives the other,

$$aa = b , \qquad bb = a .$$

Figure 8.3 shows the group table.

$$
\begin{array}{c|cc}
e & a & b \\
\hline
a & b & e \\
b & e & a
\end{array}
$$

Fig. 8.3. Group table of C_3

One realization is the set of complex numbers $\left\{1, e^{2\pi i/3}, e^{4\pi i/3}\right\}$ (the three third roots of 1) under multiplication. The number 1 serves as the identity element, and the other two numbers can respectively correspond to either a, b or b, a. Another realization is the set of rotations about a common axis by

$$\left\{0° \text{ (identity transformation)}, 120° \, (= 360°/3), 240° \, (= 2 \times 120°)\right\} ,$$

with the composition of consecutive rotation. The identity transformation of rotation by $0°$ corresponds to the identity element e, while the other two rotations can respectively correspond to either a, b or b, a.

Order 4

There are two different abstract groups of order 4. Both are Abelian. One of them has a structure similar to that of the abstract group of order 3. It is denoted C_4. Its group table is shown in Fig. 8.4. It can be realized by the set of complex numbers $\{1, i, -1, -i\}$ (the four

e	a	b	c
a	b	c	e
b	c	e	a
c	e	a	b

Fig. 8.4. Group table of C_4

e	a	b	c
a	e	c	b
b	c	e	a
c	b	a	e

Fig. 8.5. Group table of D_2

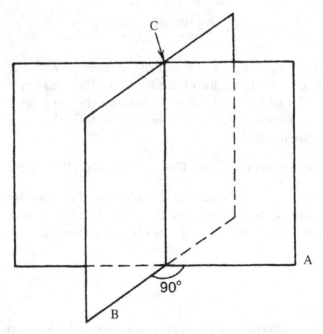

Fig. 8.6. Two perpendicular intersecting mirrors A and B and their line of intersection C

fourth roots of 1) under multiplication. Another realization is the set of rotations about a common axis by

$$\{0° \text{ (identity)}, 90° \,(= 360°/4), 180° \,(= 2 \times 90°), 270° \,(= 3 \times 90°)\} \,,$$

with consecutive rotation as composition.

The other group of order 4, denoted D_2, has the group table of Fig. 8.5. In this group each element is its own inverse,

$$aa = bb = cc = e \,.$$

To obtain a realization we imagine two perpendicular intersecting two-sided mirrors A and B and their line of intersection C, as in Fig. 8.6. The results of the transformations of reflection through each mirror and rotation by 180° about their line of intersection are shown in cross section in Fig. 8.7. The set of transformations consisting of those transformations and the identity transformation, with consecutive transformation as composition, forms a group and is a realization of D_2.

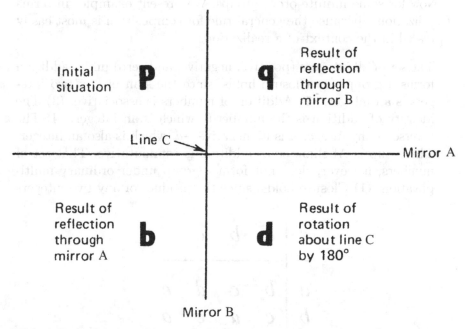

Fig. 8.7. Reflections in perpendicular intersecting mirrors A and B and rotation by 180° about their line of intersection C, shown in cross section

Order 5

There is only a single abstract group of order 5, denoted C_5, and it is Abelian. Its structure is similar to that of C_4. Figure 8.8 displays its group table. One realization is the set of five fifth roots of 1 under multiplication. Another is the set of rotations

$$\left\{0°, 72°(= 360°/5), 144°(= 2 \times 72°), 216°(= 3 \times 72°), 288°(= 4 \times 72°)\right\},$$

with consecutive rotation as composition.

All groups of orders 1 to 5 are Abelian. One might be tempted to guess by induction that all finite-order groups are Abelian. (That is the kind of 'induction' by which one 'proves' that 60 is divisible by all natural numbers: it is divisible by 2, by 3, by 4, by 5, by 6, and so on.) That is false. It turns out that the lowest-order non-Abelian group is one of the two order-6 groups, which is denoted D_3. To dispel any lingering impression that all finite-order groups are Abelian, its group table is displayed in Fig. 8.9.

Now for some infinite-order groups. We present examples in terms of realizations, because the general rule for composition is most easily expressed in the context of a realization.

1. The set of all integers (positive, negative, and zero) under addition forms a group. (1) Closure holds, since the sum of any two integers is an integer. (2) Addition of numbers is associative. (3) The identity of addition is the number 0, which is an integer. (4) The inverse of any integer a is its negative $-a$, which is also an integer. The group is Abelian, since addition is commutative. That set of numbers, however, does not form a group under ordinary multiplication. (1) Closure holds, since the product of any two integers

e	a	b	c	d
a	b	c	d	e
b	c	d	e	a
c	d	e	a	b
d	e	a	b	c

Fig. 8.8. Group table of C_5

$$
\begin{array}{c|cccccc}
 & e & a & b & c & d & f \\
\hline
a & b & e & f & c & d \\
b & e & a & d & f & c \\
c & d & f & e & a & b \\
d & f & c & b & e & a \\
f & c & d & a & b & e \\
\end{array}
$$

Fig. 8.9. Group table of D_3

is an integer. (2) Multiplication of numbers is associative. (3) The identity of multiplication is the number 1, which is an integer. But it is property 4 that is not satisfied, since the multiplicative inverse of any integer a, which is its reciprocal $1/a$, is not in general an integer; for $a = 0$ it is not even defined.

2. The set of all nonzero rational numbers, numbers expressible as a ratio of nonzero integers, does form a group under multiplication. (1) The product of any two rational numbers is rational. (2) Multiplication is associative and even commutative, so the group is Abelian. (3) The identity of multiplication, the number 1, is rational. (4) The multiplicative inverse of any nonzero rational number a is its reciprocal $1/a$, which is also rational, since if a is the ratio of two integers, so is $1/a$. This set of numbers together with the number 0 forms a group under addition.

3. The set of all $n \times n$ matrices forms a group under matrix addition. (1) The sum of any two $n \times n$ matrices is an $n \times n$ matrix. (2) Matrix addition is associative. It is also commutative, so the group is Abelian. (3) The identity of matrix addition is the $n \times n$ null matrix, the $n \times n$ matrix whose elements are all 0, which is a member of the set. (4) The inverse of any $n \times n$ matrix is its negative, which is also an $n \times n$ matrix. The same set of matrices fails to form a group under matrix multiplication. It runs into trouble with property 4.

4. However, the set of all nonsingular $n \times n$ matrices, where a nonsingular matrix is one with nonvanishing determinant, does form a group under matrix multiplication. (1) The product of any two nonsingular $n \times n$ matrices is a nonsingular $n \times n$ matrix. (2) Matrix

multiplication is associative. It is not commutative, so the group is non-Abelian (except for $n = 1$). (3) The identity of matrix multiplication is the $n \times n$ unit matrix, the $n \times n$ matrix whose diagonal elements all equal 1 and the rest of whose elements are all 0, which is nonsingular. (4) The inverse of any nonsingular $n \times n$ matrix is its matrix inverse, which exists and is a nonsingular $n \times n$ matrix. This set of matrices does not form a group under matrix addition, since it does not fulfill properties 1 and 3.

5. The set of all real orthogonal $n \times n$ matrices under matrix multiplication forms a group. A real orthogonal matrix is a matrix whose elements are all real and whose transpose, which is the matrix obtained by reflecting the matrix's elements through its diagonal, i.e., by interchanging rows and columns, is its matrix inverse. (1) The matrix product of any two real $n \times n$ orthogonal matrices is a real orthogonal $n \times n$ matrix. (2) Matrix multiplication is associative. (3) The $n \times n$ unit matrix, the identity of matrix multiplication, is real and orthogonal. (4) The inverse of any real orthogonal $n \times n$ matrix is its matrix inverse, which exists and is also a real orthogonal $n \times n$ matrix. This group is non-Abelian for $n > 2$.

6. The set of all spatial displacements (some call them translations) in a common direction forms a group with the composition of consecutive displacement. (1) The composition of any two displacements in a common direction is a displacement in the same direction. In fact, the composition of displacement by a meters and displacement by b meters in a common direction, where a and b are any two real numbers and the displacements are performed in either order, is displacement by $(a + b)$ meters in the same direction. We see that the group is Abelian, since addition of numbers is commutative. (2) Composition by consecutive displacement is associative. That might or might not be clear. In Sect. 10.2, we will see that composition of transformations by consecutive application is always associative. In the present case the composition is seen to be associative, as the addition of numbers is associative. (3) The identity transformation, displacement by 0 meters, is a member of the set and serves as the identity of the group. The identity transformation is the identity of any group of transformations with consecutive application as composition. That will be discussed in detail in Sect. 10.2. (4) For any displacement the inverse is also a displacement in the same direction. It is displacement by the same distance, in the same direction, but in the opposite sense. In other

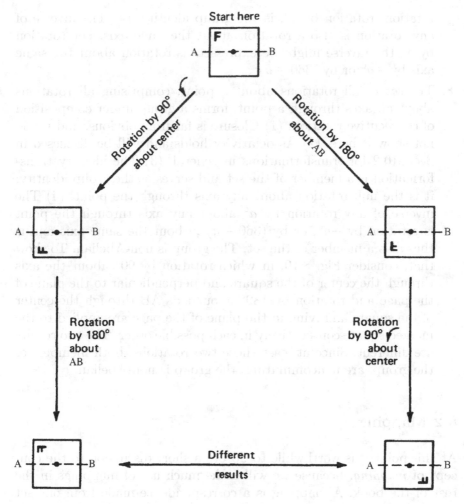

Fig. 8.10. Example of noncommuting rotations

words, the inverse of displacement by a meters is displacement by $-a$ meters in the same direction.

7. The set of all rotations about a common axis, with composition of consecutive rotation, forms a group. (1) The composition of any two rotations about the same axis is a rotation about the same axis. Rotation by $a°$ followed by rotation by $b°$ about the same axis, or vice versa, results in total rotation by $(a + b)°$ about the same axis. The group is clearly Abelian. (2) Composition of consecutive rotation is associative. That is discussed in Sect. 10.2. (3) The identity transformation is a member of the set, since the null

rotation, rotation by $0°$, is the group identity. (4) The inverse of any rotation is also a rotation about the same axis. For rotation by $a°$ the inverse might be expressed as rotation about the same axis by $-a°$ or by $(360 - a)°$.

8. The set of all rotations about a point, comprising all rotations about all axes through a point, forms a group under composition of consecutive rotation. (1) Closure is far from obvious, and we do not show it here. (2) Associativity holds and will be discussed in Sect. 10.2 for transformations in general. (3) The identity transformation is a member of the set and serves as the group identity; it is the null rotation about any axis through the point. (4) The inverse of any rotation by $a°$ about any axis through the point is rotation by $-a°$, or by $(360 - a)°$, about the same axis and is thus also a member of the set. The group is non-Abelian. To show that, consider Fig. 8.10, in which rotation by $90°$ about the axis through the center of the square and perpendicular to the plane of the page and rotation by $180°$ about axis AB through the center of the square and lying in the plane of the page are applied to the marked square consecutively in each possible order. The two results are different. Since at least those two rotations, both members of the group, are noncommuting, the group is non-Abelian.

8.2 Mapping

At this point it is worthwhile to devote a short discussion to the concept of *mapping*, because we will make much use of mappings in the rest of the book. A mapping is a correspondence made from one set to another (or to the same set). The sets may or may not be groups. A mapping from set A to set B puts every element of A in correspondence with some element of B, as in Fig. 8.11. A mapping is denoted $A \to B$. If the mapping is given a name, say mapping M from A to B, it is denoted $A \xrightarrow{M} B$ or $M : A \longrightarrow B$.

An element of set B that is in correspondence with an element of set A is called the *image* of the element of A. The element of A is called an *object* of the element of B. A mapping $A \to B$ such than an element of B may be the image of more than one element of A is called a *many-to-one mapping*. An example is shown in Fig. 8.11. If, however, different elements of A are always in correspondence with different elements of B, i.e., if every image in B is the image of a unique object in A, the mapping is called *one-to-one*. (A one-to-one mapping is also called an

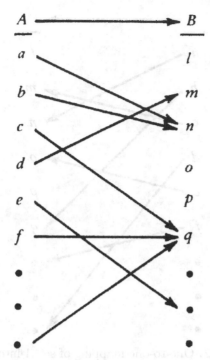

Fig. 8.11. Many-to-one mapping of set A into set B

injective mapping, in case you meet the term somewhere.) Figure 8.12 illustrates a one-to-one mapping.

If not all elements of set B are necessarily images under the mapping, it is called a mapping of A *into* B. Figures 8.11 and 8.12 illustrate such mappings. If, however, every element of B is the image of at least one element of A, the mapping is from A *onto* B, as in Fig. 8.13. (An onto mapping is also called a *surjective* mapping.)

A mapping may also be denoted in terms of the elements, as $a \longmapsto b$ or $a \overset{M}{\longmapsto} b$, where a and b represent elements of sets A and B, respectively. We use that notation extensively. Another notation, using the notation commonly used for mathematical functions, is $b = M(a)$. We also use this notation, but not in Chaps. 8 and 9.

A mapping may be from a set to itself, i.e., sets A and B may be the same set. Familiar examples of such mappings are real functions,

$$x \longmapsto y = f(x) \, ,$$

where the set is the set of all real numbers. The function $y = x^2$ is a many-to-one (actually two-to-one except for $y = 0$) mapping of the

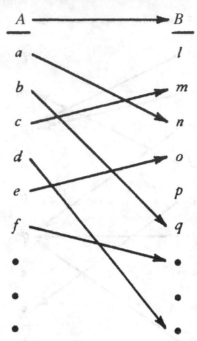

Fig. 8.12. One-to-one mapping of set A into set B

reals into the reals (only the nonnegative numbers serve as images). The function $y = \sin x$ is an infinity-to-one mapping of the reals into the reals (only numbers whose absolute values are not larger than 1 serve as images). The function $y = 3x - 7$ is a one-to-one mapping from the reals onto the reals.

A one-to-one mapping from set A onto set B may be inverted, simply by reversing the sense of the correspondence arrows, to obtain a one-to-one mapping from B onto A. The latter mapping is called the *inverse* of the former. All elements of B are objects under the inverse mapping, because all elements of B are images for the direct mapping. The inverse mapping is one-to-one, since the direct mapping is one-to-one. And the inverse mapping is onto, because all elements of A are objects under the direct mapping. Thus, the inverse mapping is indeed a one-to-one mapping from B onto A, as in Fig. 8.14. Its inverse is the direct mapping itself.

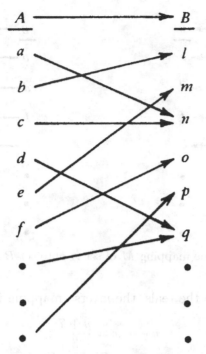

Fig. 8.13. Many-to-one mapping of set A onto set B

If the direct mapping is denoted M, its inverse is denoted M^{-1}, so that

$$A \xrightarrow{M} B , \qquad B \xrightarrow{M^{-1}} A ,$$

$$a \xmapsto{M} b , \qquad b \xmapsto{M^{-1}} a ,$$

$$b = M(a) , \qquad a = M^{-1}(b) .$$

Note that only a mapping that is both one-to-one and onto possesses an inverse; none of the other kinds of mapping we discussed is invertible. (For your information, a mapping that is both one-to-one and onto, i.e., is both injective and surjective, is called *bijective*. Thus, bijectiveness and invertibility imply one another.)

As an example of an inverse mapping, taking the one-to-one mapping

$$x \longmapsto y = 3x - 7 ,$$

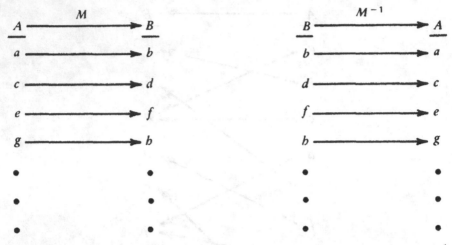

Fig. 8.14. One-to-one mapping M of set A onto set B and its inverse M^{-1}

from the reals onto the reals, the inverse mapping is

$$y \longmapsto x = \frac{y+7}{3} .$$

8.3 Isomorphism

Consider a many-to-one or one-to-one mapping of a group G onto another group G'. Then every element of G is assigned an image in G' and every element of G' is an image of at least one element of G, as in Fig. 8.15.

Now consider such a mapping that preserves structure. What we mean is that, if a and b are any two elements of G with images a' and b', respectively, in G', and if we denote by $(ab)'$ the image in G' of the composition ab in G, then

$$a'b' = (ab)' ,$$

where the composition $a'b'$ is, of course, in G'. In other words, the image of a composition is the composition of images. That can also be expressed by this diagram for all a, b in G:

$$
\begin{array}{ccc}
a & b & = c \\
\downarrow & \downarrow & \downarrow \\
a' & b' & = c'
\end{array}
$$

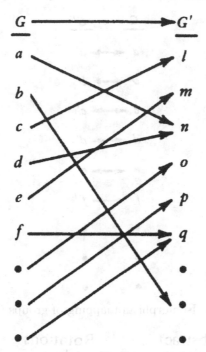

Fig. 8.15. Many-to-one mapping of group G onto group G'

Such a structure preserving mapping is called a *homomorphism*. We postpone the discussion of homomorphism in general to the next section.

If a homomorphism is one-to-one, i.e., if each element of G' is the image of exactly one element of G so that the mapping can be inverted, it is called an *isomorphism* and is denoted $G \sim G'$. Figure 8.16 shows such a mapping. And the structure preserving property of an isomorphism can be expressed by the diagram

$$
\begin{array}{ccc}
a & b = c \\
\updownarrow & \updownarrow & \updownarrow \\
a' & b' = c'
\end{array}
$$

Isomorphic groups possess the same structure. And groups having the same structure, in the sense we used previously that their group tables can be made the same by reordering rows and columns and redenoting elements, are isomorphic. So 'isomorphic' makes precise the concept of 'having the same structure' and is more general, since it is applicable to infinite-order as well as finite-order groups. Thus, all realizations

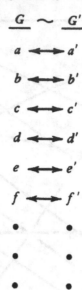

$$\underline{G} \sim \underline{G'}$$

$$a \longleftrightarrow a'$$
$$b \longleftrightarrow b'$$
$$c \longleftrightarrow c'$$
$$d \longleftrightarrow d'$$
$$e \longleftrightarrow e'$$
$$f \longleftrightarrow f'$$

Fig. 8.16. Isomorphism mapping of groups G and G'

Abstract elements		Rotations by		Numbers
e	$\longleftrightarrow$	$0°$	$\longleftrightarrow$	1
a	$\longleftrightarrow$	$120°$	$\longleftrightarrow$	$e^{2\pi i/3}$
b	$\longleftrightarrow$	$240°$	$\longleftrightarrow$	$e^{4\pi i/3}$

Mapping:

Composition:

e	a	b
a	b	e
b	e	a

Consecutive rotation   Multiplication

Fig. 8.17. Isomorphism of C_3 and two of its realizations

of an abstract group are isomorphic with the abstract group of which they are realizations as well as with each other.

Consider some examples of isomorphism.

1. The abstract group C_3 of order 3 and two of its realizations, by rotations and by complex numbers (see Fig. 8.17).

2. The Abelian infinite-order group of integers under addition is isomorphic with the Abelian infinite-order group of integral powers
of 2 (or of any other positive number) under multiplication. The
mapping is $n \leftrightarrow 2^n$. The identities are $0 \leftrightarrow 2^0 = 1$. Structure is
preserved by the mapping:

$$m + n = (m + n)$$
$$\updownarrow \quad \updownarrow \qquad \updownarrow$$
$$2^m \times 2^n = \quad 2^{m+n}$$

3. The Abelian infinite-order group of all real numbers under addition
is isomorphic with the Abelian infinite-order group of all displacements in a common direction with the composition of consecutive
displacement (see Fig. 8.18 for structure preservation).

Here and in the following we adopt the convention that transformations
are read from right to left. The reason is made clear in Sect. 10.2. For
the time being we use "+" to denote composition of transformations
by consecutive application.

4. The Abelian infinite-order group of all rotations about a common
axis, with composition of consecutive rotation, is isomorphic with
the Abelian infinite-order group of all unimodular (determinant =
1) real orthogonal 2×2 matrices under matrix multiplication. Both
are isomorphic with the group of unimodular (absolute value = 1)
complex numbers under multiplication.

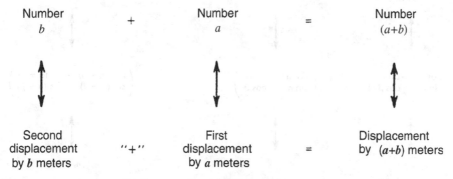

Number b	+	Number a	=	Number $(a+b)$
Second displacement by b meters	"+"	First displacement by a meters	=	Displacement by $(a+b)$ meters

Fig. 8.18. Structure preservation for the isomorphism of the group of real
numbers under addition with the group of displacements in a common direction (with composition of consecutive displacement, denoted "+")

Note that the most general 2×2 matrix that is unimodular, real, and orthogonal can be written as

$$\begin{pmatrix} \cos \alpha & -\sin \alpha \\ \sin \alpha & \cos \alpha \end{pmatrix},$$

with α real. The most general unimodular complex number is $e^{i\alpha}$ with α real. The mapping for those isomorphisms is

$$(\text{rotation by angle } \alpha) \longleftrightarrow \begin{pmatrix} \cos \alpha & -\sin \alpha \\ \sin \alpha & \cos \alpha \end{pmatrix} \longleftrightarrow e^{i\alpha}.$$

The identities are

$$(\text{rotation by angle } 0) \longleftrightarrow \begin{pmatrix} 1 & 0 \\ 0 & 1 \end{pmatrix} \longleftrightarrow e^0 = 1.$$

For structure preservation see Fig. 8.19.

5. The non-Abelian infinite-order group of all rotations about a common point under consecutive rotation is isomorphic with the non-Abelian infinite-order group of unimodular real orthogonal 3×3 matrices under matrix multiplication. The mapping is (rotation by angle α about axis with direction cosines (λ, μ, ν)) $\longleftrightarrow$ (a certain unimodular real orthogonal 3×3 matrix $M(\lambda, \mu, \nu; \alpha)$ uniquely determined by (λ, μ, ν) and α in a way that we do not go into here) [48, 49]. For structure preservation see Fig. 8.20.

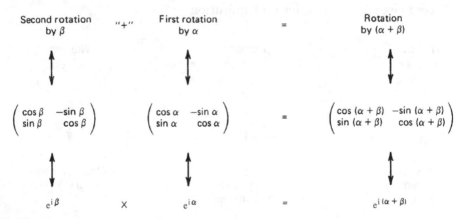

Fig. 8.19. Structure preservation for the isomorphism of the group of rotations about a common axis (with composition of consecutive rotation, denoted "+"), the group of unimodular real orthogonal 2×2 matrices under matrix multiplication, and the group of unimodular complex numbers under multiplication

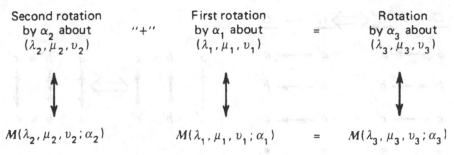

Fig. 8.20. Structure preservation for the isomorphism of the group of rotations about a common point (with composition of consecutive rotation, denoted "+") with the group of unimodular real orthogonal 3×3 matrices $M(\lambda, \mu, \nu; \alpha)$ under matrix multiplication. α is the angle of rotation about an axis with direction cosines (λ, μ, ν)

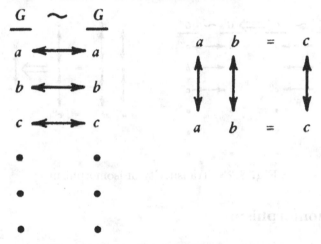

Fig. 8.21. Reflexivity of isomorphism

Referring back to Sect. 4.2, note that isomorphism is an equivalence relation among groups. It possesses the three defining properties of an equivalence relation: (1) Isomorphism is reflexive, because every group is isomorphic with itself. The mapping is simply that every element is its own image (see Fig. 8.21). (2) Isomorphism is symmetric, since the mapping is one-to-one and onto and thus invertible (see Fig. 8.22). (3) Isomorphism is transitive. If one group is isomorphic with a second and the second with a third, then the first is isomorphic with the third. Figure 8.23 is worth a hundred words.

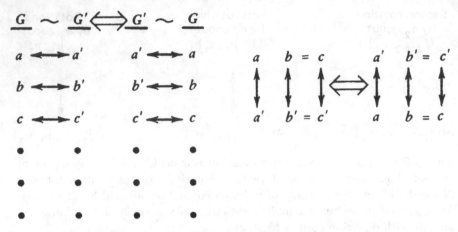

Fig. 8.22. Symmetry of isomorphism

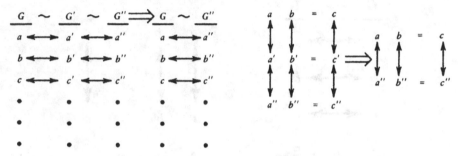

Fig. 8.23. Transitivity of isomorphism

8.4 Homomorphism

We now return to the discussion of *homomorphism* (of which isomorphism is a special case) in general, which we started and abandoned in the preceding section. Such a mapping is not invertible (unless it is an isomorphism and only then). Homomorphism is denoted $G \to G'$, where the mapping is many-to-one from group G to group G', as in Fig. 8.24. As was pointed out in the preceding section, the structure preserving property of homomorphism can be expressed by

$$a'b' = (ab)' ,$$

or by the diagram

$$
\begin{array}{ccc}
a & b & = c \\
\downarrow & \downarrow & \downarrow \\
a' & b' & = c'
\end{array}
$$

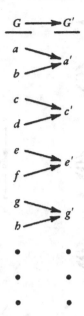

Fig. 8.24. Homomorphism mapping of group G to group G'

where a, b, ab, c are elements of G and a', b', $(ab)'$, c' their respective images in G'.

Although homomorphism preserves the group structure in the sense just shown, homomorphic groups do not possess the same structure, unless they are isomorphic. For example, non-Abelian groups may be homomorphic to Abelian groups (the opposite is impossible, however). And while isomorphic groups have the same order, infinite-order groups may be homomorphic to finite-order groups, and groups of different finite orders may be homomorphic. I will present examples shortly. Homomorphism is not an equivalence relation, since it is not symmetric. It is reflexive, however, with isomorphism as a special case of homomorphism (see Fig. 8.21). And it is transitive (see Fig. 8.25).

In a homomorphism $G \to G'$ the set of all elements of G whose images are the identity element of G' is called the *kernel* of the homomorphism. We will return to that in Sect. 9.1.

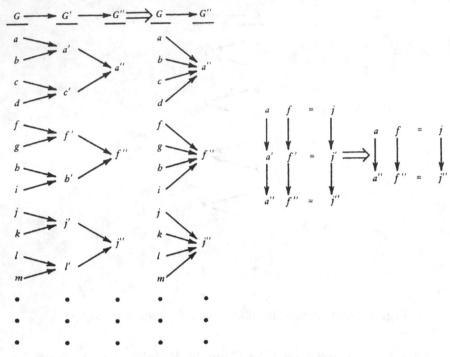

Fig. 8.25. Transitivity of homomorphism

Here are some examples of homomorphism.

1. Any group G whatsoever is trivially homomorphic to the abstract group of order 1, denoted C_1, as in Fig. 8.26. Structure is preserved:

$$
\begin{array}{ccc}
a & b = c \\
\downarrow & \downarrow \quad \downarrow \\
e' & e' = e'
\end{array}
$$

All of G serves as the kernel of such a homomorphism In terms of realizations, we can homomorphically map all the elements of any group to the number 1 (under multiplication) or to the number 0 (under addition).

2. Both abstract groups of order 4 (and any realizations thereof) are homomorphic to the abstract group of order 2, C_2 (and any of its realizations). For C_4 the mapping is shown in Fig. 8.27. For D_2 there are three possible mappings, shown in Fig. 8.28. Structure preservation can be checked with the help of the group tables shown in Sect. 8.1. The kernel of the first homomorphism (Fig. 8.27) is $\{e, b\}$.

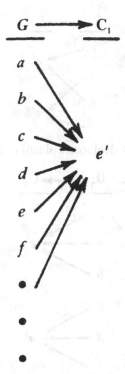

Fig. 8.26. Trivial homomorphism of any group G to the order-1 group C_1

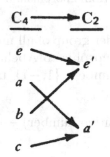

Fig. 8.27. Homomorphism of C_4 to C_2

The three possible kernels of the second homomorphism (Fig. 8.28) are $\{e, a\}$, $\{e, b\}$, and $\{e, c\}$.

3. The non-Abelian order-6 group D_3 is homomorphic to Abelian C_2. The mapping is shown in Fig. 8.29. Preservation of structure is checked with the group tables in Sect. 8.1. The kernel of the homomorphism is $\{e, a, b\}$.

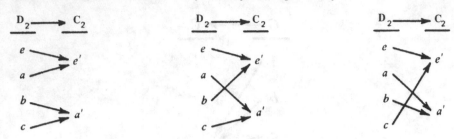

Fig. 8.28. Homomorphism of D_2 to C_2

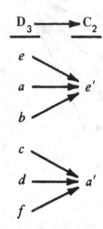

Fig. 8.29. Homomorphism of D_3 to C_2

4. The Abelian infinite-order group of all nonzero real numbers under multiplication is homomorphic to Abelian C_2 and its realization, for example, by the numbers $\{1, -1\}$ under multiplication. The mapping is

$$\text{(any positive number)} \longrightarrow e \longleftrightarrow 1 \,,$$

$$\text{(any negative number)} \longrightarrow a \longleftrightarrow -1 \,.$$

Structure preservation is seen as follows:

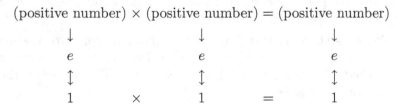

(positive number) × (negative number) = (negative number)

$$\downarrow \qquad\qquad \downarrow \qquad\qquad \downarrow$$

$$e \qquad\qquad\quad a \qquad\qquad\quad a$$

$$\updownarrow \qquad\qquad \updownarrow \qquad\qquad \updownarrow$$

$$1 \quad\times\quad (-1) \quad=\quad (-1)$$

(negative number) × (positive number) = (negative number)

$$\downarrow \qquad\qquad \downarrow \qquad\qquad \downarrow$$

$$a \qquad\qquad\quad e \qquad\qquad\quad a$$

$$\updownarrow \qquad\qquad \updownarrow \qquad\qquad \updownarrow$$

$$(-1) \quad\times\quad 1 \quad=\quad (-1)$$

(negative number) × (negative number) = (positive number)

$$\downarrow \qquad\qquad \downarrow \qquad\qquad \downarrow$$

$$a \qquad\qquad\quad a \qquad\qquad\quad e$$

$$\updownarrow \qquad\qquad \updownarrow \qquad\qquad \updownarrow$$

$$(-1) \quad\times\quad (-1) \quad=\quad 1$$

The set of all positive numbers forms the kernel of the homomorphism.

5. The non-Abelian (for $n > 1$) group of all nonsingular (having nonvanishing determinant) complex $n \times n$ matrices under matrix multiplication is homomorphic to the Abelian group of nonzero complex numbers under multiplication. Each matrix A is mapped to the complex number that is the value of its determinant, $A \to |A|$ (so that $A^{-1} \to 1/|A|$). Structure preservation follows from the fact that the determinant of a matrix product is the product of the individual determinants:

$$A \qquad B = (AB)$$

$$\downarrow \quad \downarrow \quad \downarrow$$

$$|A| \times |B| = |AB|$$

The unimodular $n \times n$ matrices make up the kernel of the homomorphism, since they are all the $n \times n$ matrices whose image is 1, the identity of number multiplication.

8.5 Subgroup

A *subset* of a set S is any set all of whose elements are also elements of S. For example, the set of positive integers is a subset of the set of integers, or the set of real matrices is a subset of the set of complex matrices. Note that according to the definition any set forms a subset of itself. If set T is a subset of set S, the relation is denoted $T \subset S$ and $S \supset T$, where $\subset$ and $\supset$ denote the relation of *inclusion*. (Sometimes $\subset$ and $\supset$ are used to denote only strict inclusion, where the sets may not be the same, so that there is at least one element in the including set that is not a member of the subset. Inclusion that allows the subset to coincide with the including set is then denoted $\subseteq$ and $\supseteq$. However, we do not use that convention.)

If a subset of the elements of a group G itself forms a group with respect to the composition law of G, it is called a *subgroup* of G. If H is a subgroup of G, the relation is denoted $H \subset G$ and $G \supset H$, using the same notation as for set inclusion.

Every group G possesses two trivial subgroups: (1) the group of order 1 consisting of the identity element of G and (2) the group G itself. Any subgroup of a group except the group itself is a *proper subgroup*.

It can be shown, and we will show in Sect. 9.2, that if a finite-order group of order n includes a subgroup of order m, then m is a divisor of n, i.e., $n = ms$ for some integer s. From that theorem it follows that groups of prime order possess no proper subgroups except the trivial subgroup of order 1. Thus, the groups of orders 2, 3, and 5, which were presented in Sect. 8.1, can have no nontrivial proper subgroups, as an examination of their group tables will confirm. In any case the group of order 2 is of too low an order to have nontrivial proper subgroups.

If G' is a subgroup of G and G'' is a subgroup of G', then clearly G'' is also a subgroup of G.

Consider the following examples of subgroups.

1. In the order-4 group C_4 (rotations by $\{0°, 90°, 180°, 270°\}$) the elements $\{e, b\}$, as shown in Fig. 8.4 of Sect. 8.1, (rotations by $\{0°, 180°\}$) form an order-2 subgroup, C_2. That is the only non-trivial subgroup of C_4.

2. The order-4 group D_2, as shown in Fig. 8.5 of Sect. 8.1, includes three order-2 subgroups: $\{e, a\}$, $\{e, b\}$, and $\{e, c\}$. According to the above theorem, the order-4 subgroups, C_4 and D_2, cannot possess

order-3 subgroups, since three is not a divisor of four. So we have found all their nontrivial subgroups.

3. An order-6 group can have nontrivial subgroups of orders 2 and 3 only. In fact, the non-Abelian order-6 group D_3, shown in Fig. 8.9 in Sect. 8.1, includes four such subgroups, one of order 3 and three of order 2: $\{e, a, b\}$, $\{e, c\}$, $\{e, d\}$, $\{e, f\}$.

4. The group of nonzero real numbers under multiplication includes various subgroups. Among those are the positive real numbers, the nonzero rational numbers, the positive rational numbers, and the integral powers of a fixed real number.

We conclude this section with a search for the nontrivial subgroups of the order-8 group D_4, whose group table is shown in Fig. 8.30. By the divisor theorem, we know we are looking only for subgroups of orders 2 and 4. Examine the diagonal of the table. Every appearance of e indicates an order-2 subgroup, since $b^2 = d^2 = f^2 = g^2 = h^2 = e$. So all the order-2 subgroups are $\{e, b\}$, $\{e, d\}$, $\{e, f\}$, $\{e, g\}$, $\{e, h\}$.

Look at powers of a: $a^2 = b$, $a^3 = ab = c$, $a^4 = ac = e$. Thus, we found an order-4 subgroup $\{e, a, b, c\}$, which is C_4. Attempting to apply the power trick further does not prove useful.

Let us see whether we can combine some of the order-2 subgroups to form order-4 subgroups. We see that $dg = gd = b$. Also, $fh = hf = b$. None of the other compositions are particularly useful. So we have two more order-4 subgroups, $\{e, b, d, g\}$ and $\{e, b, f, h\}$, which are both D_2. That exhausts the nontrivial subgroups of D_4.

e	a	b	c	d	f	g	h
a	b	c	e	f	g	h	d
b	c	e	a	g	h	d	f
c	e	a	b	h	d	f	g
d	h	g	f	e	c	b	a
f	d	h	g	a	e	c	b
g	f	d	h	b	a	e	c
h	g	f	d	c	b	a	e

Fig. 8.30. Group table of D_4

8.6 Summary

This chapter was the start of an introduction to group theory, the mathematical language of symmetry. In it we discussed the most fundamental of the fundamental concepts and ideas of group theory. To summarize, here is a list of the more important concepts discussed in each section:

- Section 8.1: element, group, order, composition, closure, associativity, identity element, inverse element, noncommuting elements, commuting elements, Abelian (commutative) group, group structure, group table, realization.
- Section 8.2: mapping, object, image, many-to-one mapping, one-to-one mapping, into mapping, onto mapping, inverse mapping.
- Section 8.3: isomorphism.
- Section 8.4: homomorphism, kernel.
- Section 8.5: subgroup, proper subgroup.

Group Theory Continued

In this chapter we build on the foundation that was laid in the preceding chapter to achieve a reasonable introduction to group theory.

9.1 Conjugacy, Invariant Subgroup, Kernel

If for some pair of elements a and b of group G there exists a (not necessarily unique) element u of G such that

$$u^{-1}au = b \, ,$$

then a and b are called *conjugate elements* in G. Conjugacy is denoted

$$a \equiv b \, .$$

It is an equivalence relation among group elements. (We presented and discussed the concept of equivalence relation earlier in Sect. 4.2.) Here is how conjugacy fulfills the three properties of an equivalence relation:

1. Conjugacy is reflexive, since every element a is conjugate with itself, $a \equiv a$ for all a in G. That is due to the fact that $e^{-1}ae = a$.
2. Conjugacy is symmetric, since for all a and b in G if $a \equiv b$, then $b \equiv a$. That comes about because if there exists an element u in G such that $u^{-1}au = b$, then $v^{-1}bv = a$ with $v = u^{-1}$. To prove it, substitute $v = u^{-1}$ in $v^{-1}bv = a$ and see what develops. First we obtain

$$(u^{-1})^{-1}bu^{-1} = a \, .$$

Since the inverse of an inverse is the element itself, this becomes

$$ubu^{-1} = a \ .$$

Now perform the composition of each side of the equation with u on the right and with u^{-1} on the left, and for convenience interchange the sides of the equation:

$$
\begin{aligned}
u^{-1}au &= u^{-1}(ubu^{-1})u \\
&= (u^{-1}u)b(u^{-1}u) \quad \text{(by associativity)} \\
&= ebe \quad \text{(by inverse)} \\
&= e(be) \quad \text{(by associativity)} \\
&= eb \quad \text{(by identity)} \\
&= b \quad \text{(by identity)} ,
\end{aligned}
$$

which is what we started from, $u^{-1}au = b$. So symmetry is proved.

3. And finally, conjugacy is transitive, since for all a, b, c in G if $a \equiv b$ and $b \equiv c$, then $a \equiv c$. If there exist elements u and v of G such that $u^{-1}au = b$ and $v^{-1}bv = c$, then $w^{-1}aw = c$ with $w = uv$. To prove that, substitute $b = u^{-1}au$ into $c = v^{-1}bv$:

$$
\begin{aligned}
c &= v^{-1}(u^{-1}au)v \quad \text{(by substitution)} \\
&= (v^{-1}u^{-1})a(uv) \quad \text{(by associativity)} \\
&= (uv)^{-1}a(uv) \quad \text{(by inverse of product)} \\
&= w^{-1}aw \quad \text{(by putting } uv = w) ,
\end{aligned}
$$

which proves transitivity, $a \equiv c$.

Note that more than two elements might be conjugate with each other. The identity element is conjugate only with itself:

$$
\begin{aligned}
u^{-1}eu &= u^{-1}(eu) \quad \text{(by associativity)} \\
&= u^{-1}u \quad \text{(by identity)} \\
&= e \quad \text{(by inverse)} ,
\end{aligned}
$$

for any element u. In an Abelian group every element is conjugate only with itself:

$$u^{-1}au = u^{-1}(au) \quad \text{(by associativity)}$$
$$= u^{-1}(ua) \quad \text{(by commutativity)}$$
$$= (u^{-1}u)a \quad \text{(by associativity)}$$
$$= ea \quad \text{(by inverse)}$$
$$= a \quad \text{(by identity)},$$

for all elements a and u of the group.

As an example, we find the conjugate elements of the order-6 group D_3, whose group table is displayed in Fig. 8.9 in Sect. 8.1 and Fig. 9.11 in Sect. 9.8. First find all the inverses. Since $ab = ba = e$, we have $a^{-1} = b$ and $b^{-1} = a$. And since $cc = dd = ff = e$, we know that $c^{-1} = c$, $d^{-1} = d$, and $f^{-1} = f$. Now conjugate each element of the group by all the others to find conjugates. For example,

$$c^{-1}ac = c(ac) = cf = b,$$

$$d^{-1}cd = d(cd) = da = f,$$

$$f^{-1}cf = f(cf) = fb = d.$$

In that way obtain the breakdown into conjugate elements:

$$e, \quad a \equiv b, \quad c \equiv d \equiv f.$$

A subset of group elements that consists of a complete set of mutually conjugate elements is called a *conjugacy class* of the group. The identity element, since it in not conjugate with any other element, is always in a conjugacy class of its own. In an Abelian group every element is in a conjugacy class of its own. For example, the conjugacy classes of the group D_3 are $\{e\}$, $\{a, b\}$, $\{c, d, f\}$, as we just found.

At this point recall or review the concept of equivalence class, which was presented and discussed earlier in Sect. 4.2. If we have a set of elements (which might form a group) for which an equivalence relation is defined, any subset that contains a complete set of mutually equivalent elements is an equivalence class. It is easily shown that different equivalence classes in the same set cannot possess common elements. Thus, an equivalence relation brings about a decomposition of the set for which it is defined, such that every element of the set is a member of one and only one equivalence class.

Two examples of equivalence class that we have already met are conjugacy classes, where the equivalence relation is conjugacy between group elements, and classes of groups with the same structure, where the equivalence relation is isomorphism of groups in the set of all groups. For the D$_3$ example above, note how the group does indeed decompose into its three conjugacy classes.

Conversely, any decomposition of a set into subsets such that every element of the set is a member of one and only one subset defines an equivalence relation. The subsets may be declared, by fiat, equivalence classes, and the corresponding equivalence relation is simply that two elements are equivalent if and only if they belong to the same subset.

Two trivial equivalence class decompositions for any set of elements are the following. The most exclusive equivalence relation, that every element is equivalent only with itself and with no other, decomposes the set into as many equivalence classes as there are elements in the set, since every equivalence class contains but a single element. And on the other hand, the most inclusive equivalence relation, that all elements of the set are equivalent with each other, makes the whole set a single equivalence class in itself.

Let us return to conjugation. Conjugation by a single element has an interesting and useful property. Let a, a', b, b', c, c' be elements of group G such that $a \equiv a'$, $b \equiv b'$, $c \equiv c'$ with respect to conjugation by the same element u of G:

$$a' = u^{-1}au\,, \qquad b' = u^{-1}bu\,, \qquad c' = u^{-1}cu\,.$$

Now, assume $ab = c$. Compose each side of the relation with u^{-1} on the left and with u on the right. The right-hand side becomes $u^{-1}cu = c'$. The left-hand side becomes

$$u^{-1}(ab)u = u^{-1}(a(eb))u \qquad \text{(by identity)}$$
$$= u^{-1}aebu \qquad \text{(by associativity)}$$
$$= u^{-1}a(uu^{-1})bu \qquad \text{(by inverse)}$$
$$= (u^{-1}au)(u^{-1}bu) \qquad \text{(by associativity)}$$
$$= a'b'\,.$$

Thus, $a'b' = c'$, and intragroup relations are invariant under conjugation by the same element. If you happen to be familiar with the invariance of algebraic relations among matrices under similarity transformations, note the analogy.

For example, in the group D_3, whose group table is displayed in Fig. 8.9 in Sect. 8.1 and Fig. 9.11 in Sect. 9.8, we find the relation $bd = f$. Under conjugation by c the element b becomes b', where

$$b' = c^{-1}bc = cbc \qquad \text{(c is its own inverse)}$$

$$= c(bc) \qquad \text{(by associativity)}$$

$$= cd \qquad \text{(from the group table)}$$

$$= a \qquad \text{(from the group table)}.$$

Similarly,

$$d' = c^{-1}dc = f\,, \qquad f' = c^{-1}fc = d\,.$$

Invariance of intragroup relations means that $b'd' = f'$, or in the present case that $af = d$. Check in the D_3 group table to confirm that this is indeed true.

If H is a subset (not necessarily a subgroup) of group G, $g^{-1}Hg$ for some g in G denotes the subset of G consisting of elements $g^{-1}hg$, where h runs over all elements of H, i.e., $g^{-1}Hg$ is the subset conjugate with H by g. If G is Abelian, $g^{-1}Hg = H$ for all g in G.

If H is a subgroup of G, then $g^{-1}Hg$ for any g in G is also a subgroup of G. That is seen as follows:

1. Since H is a subgroup, it is closed under the composition of G. Therefore, so is $g^{-1}Hg$ closed, since, as we saw above, intragroup relations are invariant under conjugation by the same element (here g).

2. Associativity holds for $g^{-1}Hg$, since it holds for G and $g^{-1}Hg$ is a subset of G.

3. The set $g^{-1}Hg$ contains the identity, since H, being a subgroup, contains it and $g^{-1}eg = e$.

4. Any element of $g^{-1}Hg$ is expressible as $g^{-1}hg$, where h is some element of H. The inverse of $g^{-1}hg$ is $(g^{-1}hg)^{-1} = g^{-1}h^{-1}g$, which is also an element of $g^{-1}Hg$, since h^{-1} is an element of H. Thus, $g^{-1}Hg$ contains the inverses of all its elements.

So $g^{-1}Hg$ possesses the four group properties and is indeed a group if H is. And, again since intragroup relations are invariant under conjugation by the same element, conjugate subgroups are isomorphic.

For example, D_3 includes the subgroup $\{e, c\}$. The conjugate subgroup by a is $\{e, f\}$:

$$a^{-1}\{e, c\}a = b\{e, c\}a = \{bea, bca\} = \{e, f\} .$$

Both subgroups are isomorphic, since they are both of order 2 and there is only one abstract order-2 group, C_2.

If H is a subgroup of G and $g^{-1}Hg = H$ for all g in G, i.e., if H is conjugate only with itself, H is called an *invariant subgroup* (also *normal subgroup*).

As an example, the subgroup $\{e, a, b\}$ of D_3 is an invariant subgroup. Earlier in this section we found that elements a and b are conjugate with each other and with no other element, while the identity e is conjugate only with itself. Thus,

$$g^{-1}\{e, a, b\}g = \{g^{-1}eg, g^{-1}ag, g^{-1}bg\} = \{e, a, b\} ,$$

for all elements g of D_3 and the subgroup is proved invariant as claimed. If you care to, check this for $g = e$, a, b, c, d, f. You might find your result looking like $\{e, b, a\}$, but it is the same set as $\{e, a, b\}$.

Note that the trivial subgroups $\{e\}$ and G of any group G are invariant. In an Abelian group all subgroups are invariant.

Now, consider the homomorphism $G \rightarrow G'$. The subset of elements of G that is mapped to the identity element e' of G' is called the *kernel* of the homomorphism, as was mentioned in Sect. 8.4. A kernel is usually denoted K. Thus K consists of all g in G for which $g \rightarrow e'$. For any homomorphism $G \rightarrow G'$ the kernel K is an invariant subgroup of G, as we proceed to prove. First we show that K is a subgroup of G:

1. Let a and b be any pair of elements of K, so that $a \rightarrow e'$ and $b \rightarrow e'$. Denote the image of their composition $c\ (= ab)$ by c'. Preservation of structure by homomorphism requires

$$
\begin{array}{ccc}
a & b & = c \\
\downarrow & \downarrow & \downarrow \\
e' & e' & = e'
\end{array}
$$

   But $e'e' = e'$, so $c' = e'$, $c \rightarrow e'$, and $c \in K$. Thus K is closed under the composition of G.

2. Associativity holds for K, since it holds for G and K is a subset of G.

3. Is e an element of K? The identity element e obeys $ea = a$ for all a in G. Let $e \to x'$ and $a \to a'$ under the homomorphism. Since a runs over all elements of G, a' runs over all elements of G'. Preservation of structure requires

$$e\ a = a$$
$$\downarrow\downarrow \quad \downarrow$$
$$x'\ a' = a'$$

Thus, $x'a' = a'$, for all a' in G', which means that $x' = e'$, since the identity is unique. So $e \to e'$, and e is indeed an element of K.

4. Let a be any element of K, so that $a \to e'$, and denote the image of its inverse by $(a^{-1})'$. By preservation of structure

$$a\ a^{-1} = e$$
$$\downarrow \quad \downarrow \qquad \downarrow$$
$$e'\ (a^{-1})' = e'$$

Thus, $(a^{-1})' = e'$, and a^{-1} is an element of K. So K contains the inverses of all its elements.

Since K is endowed with all four group properties, it is a group, a subgroup of G.

We now prove that K is an *invariant* subgroup of G. Let a be any element of K so that $a \to e'$, and form the conjugation $g^{-1}ag = b$ for all elements g of G. Denote by b', g', and $(g^{-1})'$ the images of b, g, and g^{-1}, respectively. Preservation of structure gives us $(g^{-1})' = g'^{-1}$ and

$$g^{-1}\ a\ g = b$$
$$\downarrow \quad \downarrow\ \downarrow \qquad \downarrow$$
$$g'^{-1}\ e'\ g' = b'$$

from which it follows that $b' = e'$ and $b \in K$. Thus, $g^{-1}ag$ is an element of K for all a in K and all g in G. So $g^{-1}Kg = K$ for all g in G, and K is an invariant subgroup of G as claimed.

If the kernel of a homomorphism $G \to G'$ consists only of the identity element e, the mapping is, as we will see in Sect. 9.4, one-to-one, and the homomorphism is actually an isomorphism. If the kernel is the group G itself, the homomorphism is the trivial one $g \to e'$ for all g in G, where G' is the order-1 group.

Consider a few examples of homomorphism kernels as invariant subgroups:

1. For the homomorphism $D_3 \rightarrow C_2$, presented as an example in Sect. 8.4, the kernel is $\{e, a, b\}$. It was shown earlier in the present section that it is an invariant subgroup of D_3.

2. Consider the homomorphism, also presented as an example in Sect. 8.4, of the group of nonsingular (having nonzero determinant) complex $n \times n$ matrices under matrix multiplication to the group of nonzero complex numbers under multiplication, where each matrix is mapped to the complex number that is the value of its determinant, $A \rightarrow |A|$. The kernel of the homomorphism is the subset consisting of all unimodular $n \times n$ matrices, i.e., all complex $n \times n$ matrices whose determinant has the value 1. Now we show that the kernel is a subgroup:

   (1) The product of unimodular $n \times n$ matrices is a unimodular $n \times n$ matrix. That follows from the fact that the determinant of a matrix product equals the product of the individual determinants, i.e., $|AB| = |A||B|$ for any matrices A and B. So the kernel is closed under matrix multiplication.

   (2) Matrix multiplication is associative.

   (3) The unit $n \times n$ matrix, the identity of matrix multiplication, is unimodular.

   (4) The inverse of any unimodular $n \times n$ matrix exists, because every nonsingular matrix is invertible. The inverse is an $n \times n$ matrix and is unimodular, since if $|A| = 1$, then $|A^{-1}| = 1/|A| = 1$.

Having the four group properties, this kernel is confirmed to be a subgroup (which was proved in general earlier in this section).

Now check that it is an *invariant* subgroup (which it must be, as was proved in general). Let A be any nonsingular $n \times n$ matrix and U any unimodular $n \times n$ matrix. Form the conjugate element of U by A, $A^{-1}UA$ (which is just the similarity transformation of U by A). The value of its determinant is

$$|A^{-1}UA| = |A^{-1}||U||A|$$
$$= (1/|A|)|U||A|$$
$$= |U|$$
$$= 1 .$$

So $A^{-1}UA$ is always unimodular and an element of the kernel, which is thus confirmed to be an invariant subgroup.

3. As we saw in Sect. 8.4, the group of all nonzero real numbers under multiplication is homomorphic to the group of numbers $\{1, -1\}$ under multiplication, with the mapping

$$(\text{any positive number}) \longrightarrow 1 \, ,$$

$$(\text{any negative number}) \longrightarrow -1 \, .$$

Since the number 1 serves as identity in $\{1, -1\}$, the kernel of the homomorphism is the subset consisting of all positive real numbers, which is indeed a group under multiplication. Moreover, it is trivially invariant, as multiplication of numbers is commutative.

9.2 Coset Decomposition

Let H be any proper subgroup of group G (i.e., any subgroup of G that is not G itself). Form the set aH, where a is any element of G that is not a member of H. That is the set of elements ah, where h runs over all the elements of H. Such a set is called a *left coset* of H in G. Similarly, Ha is called a *right coset*. Coset aH is not a subgroup of G, since it does not contain the identity element e. If it did contain e, H would have to contain a^{-1}. Since H is a subgroup, it would then have to contain also the inverse of a^{-1}, which is a. But by assumption H does not contain a. So aH is not a subgroup.

If H is of order m, then aH also contains m elements. The coset certainly contains no more than m elements and could contain less only if $ah_1 = ah_2$ for some distinct pair of elements h_1 and h_2 of H. But $ah_1 = ah_2$ implies $h_1 = h_2$ by composition with a^{-1} on the left, which is a contradiction. Therefore, aH contains exactly m elements.

In addition, no element of aH is also an element of H. Otherwise we would have $ah_1 = h_2$ for some pair of elements h_1, h_2 of H. That implies $a = h_2 h_1^{-1}$ by composition with h_1^{-1} on the right. Since H is a subgroup and is closed under composition, $a = h_2 h_1^{-1}$ means that a is an element of H, contrary to assumption. So H and aH have no element in common.

The *union* of a number of given sets is the set of all distinct elements of the given sets. The symbol $\cup$ is used to denote union of sets. Form the union of H and aH, $H \cup aH$. The set $H \cup aH$ might or might not exhaust all the elements of G. If it does not, take any element b of G that is not a member of the set $H \cup aH$, i.e., b is an element neither of H nor of aH, and form the left coset bH. By reasoning similar to the above, bH has exactly m elements and has no element in common with H or with aH. Now form the union $H \cup aH \cup bH$. If that set does not exhaust G, continue forming left cosets in the same manner. If G is of finite order n, the procedure will eventually exhaust it:

$$G = H \cup aH \cup bH \cup \ldots \cup kH \ .$$

Since H and each of its cosets contain m elements per set and the sets have no element in common, we obtain the result that $n = ms$ for some integer s, i.e., m is a divisor of n, as mentioned in Sect. 8.5. This result is known as Lagrange's theorem (Joseph Louis Lagrange, French mathematician and astronomer, 1736–1813). Note that the proof can be performed just as well with right cosets.

The decomposition of a group G into cosets of any subgroup will always be the same for left cosets as for right cosets if G is Abelian. If G is non-Abelian, the decompositions might differ, although not necessarily. The decomposition of G into left (or right) cosets of subgroup H is unique. By that we mean that, if we form a left coset lH of H with any element l of G that is not a member of H and is not among the elements $a, b, \ldots, k$ used in the decomposition

$$G = H \cup aH \cup bH \cup \ldots \cup kH \ ,$$

the coset lH will necessarily be one of the cosets aH, bH, $\ldots$, kH. To see that, note that l must be a member of some one of those cosets, say aH. Since all elements of aH are of the form ah, where h runs over all elements of H, then $l = ah_1$ for some element h_1 of H. Thus, by composition with h_1^{-1} on the right, $a = lh_1^{-1}$ and all elements of aH are of the form

$$ah = \left(lh_1^{-1}\right)h = l\left(h_1^{-1}h\right) \ ,$$

where h, and consequently also $h_1^{-1}h$, runs over all elements of H. Therefore, $aH = lH$. So the same decomposition of G into left (or right) cosets of subgroup H is obtained whatever elements are used to form the cosets.

Examples of decomposition into cosets follow. Refer to Sect. 8.1 for group tables of the groups involved.

1. The order-4 group C_4 includes a single nontrivial proper subgroup, $H = \{e, b\}$. Its left coset formed with a is

$$aH = \{ae, ab\} = \{a, c\} \ .$$

Its left coset formed with c is

$$cH = \{ce, cb\} = \{c, a\} = aH \ .$$

So the unique decomposition of the group into left cosets of H is

$$G = \{e, b\} \cup \{a, c\} \ .$$

Since the group is Abelian, decomposition into right cosets of H is the same, which is obvious in this example, as there is only a single possibility for any coset of H.

2. The order-4 group D_2 possesses three nontrivial proper subgroups, which are $\{e, a\}$, $\{e, b\}$, and $\{e, c\}$. So we have three different decompositions into left cosets:

$$H = \{e, a\} \ , \qquad bH = cH = \{b, c\} \ , \qquad G = \{e, a\} \cup \{b, c\} \ ,$$

$$H = \{e, b\} \ , \qquad cH = aH = \{a, c\} \ , \qquad G = \{e, b\} \cup \{a, c\} \ ,$$

$$H = \{e, c\} \ , \qquad aH = bH = \{a, b\} \ , \qquad G = \{e, c\} \cup \{a, b\} \ .$$

Right and left cosets of the same subgroup are the same in this example, since D_2 is Abelian.

3. The non-Abelian order-6 group D_3, whose group table is shown also in Fig. 9.11 in Sect. 9.8, includes the nontrivial proper subgroups $\{e, a, b\}$, $\{e, c\}$, $\{e, d\}$, and $\{e, f\}$. So we have four decompositions into left cosets:

$$H = \{e, a, b\} \ , \quad cH = \{c, d, f\} \ , \quad G = \{e, a, b\} \cup \{c, d, f\} \ ,$$

$$H = \{e, c\} \ , \quad aH = \{a, f\} \ , \quad bH = \{b, d\} \ ,$$

$$G = \{e, c\} \cup \{a, f\} \cup \{b, d\} \ ,$$

$$H = \{e, d\} \ , \quad aH = \{a, c\} \ , \quad bH = \{b, f\} \ ,$$

$$G = \{e, d\} \cup \{a, c\} \cup \{b, f\} \ ,$$

$$H = \{e, f\} , \quad aH = \{a, d\} , \quad bH = \{b, c\} ,$$

$$G = \{e, f\} \cup \{a, d\} \cup \{b, c\} .$$

Since the group is non-Abelian, its decompositions into right cosets might be different from its decompositions into left cosets. Its four decompositions into right cosets are

$$H = \{e, a, b\} , \quad Hc = \{c, d, f\} , \quad G = \{e, a, b\} \cup \{c, d, f\} ,$$

$$H = \{e, c\} , \quad Ha = \{a, d\} , \quad Hb = \{b, f\} ,$$

$$G = \{e, c\} \cup \{a, d\} \cup \{b, f\} ,$$

$$H = \{e, d\} , \quad Ha = \{a, f\} , \quad Hb = \{b, c\} ,$$

$$G = \{e, d\} \cup \{a, f\} \cup \{b, c\} ,$$

$$H = \{e, f\} , \quad Ha = \{a, c\} , \quad Hb = \{b, d\} ,$$

$$G = \{e, f\} \cup \{a, c\} \cup \{b, d\} .$$

Thus, decomposition into cosets of subgroup $\{e, a, b\}$ is the same for left and for right, whereas decomposition into cosets of any of the order-2 subgroups is different for left and for right.

Let H be any invariant proper subgroup of group G, so that $g^{-1}Hg = H$ for all g in G. By composition with g on the left, the relation can be put in the form $Hg = gH$ for all g in G, with the result that left and right cosets of any invariant subgroup are the same. But be careful! The relation $gH = Hg$ for all g in G does not mean that all elements of G commute with all elements of H. Rather, the set of elements hg, where H runs over all elements of H, is the same as the set of elements gh, where h runs over all elements of H, for all elements g of G. Thus, decomposition of a group into cosets of an invariant subgroup is the same whether left or right cosets are used. We saw that in the example of decomposition of D_3 into cosets of its invariant subgroup $\{e, a, b\}$. In an Abelian group all subgroups are invariant, so left-coset and right-coset decompositions are the same for any subgroup.

9.3 Factor Group

In Sect. 9.2 we found that decomposition of a group into left cosets of an invariant subgroup is the same as decomposition into right cosets of the same invariant subgroup. That fact has far-reaching consequences, as we will see in the present section and the next.

We now define a composition of cosets of an invariant subgroup H of group G. It is denoted $(aH)(bH)$ and is defined as the set of all elements $h'h''$, where h' runs over all elements of coset aH and h'' independently runs over all elements of coset bH. That can also be expressed as the set of elements $(ah_1)(bh_2) = ah_1bh_2$, where h_1 and h_2 independently run over all elements of H. Using associativity and equality of left and right cosets of the invariant subgroup H, we obtain the following string of equalities for the composition of cosets aH and bH, as defined just now:

$$(aH)(bH) = a(Hb)H$$
$$= a(bH)H$$
$$= (ab)(HH)$$
$$= (ab)H \, .$$

The last equality follows from $HH = H$, since HH is the set of elements h_1h_2, where h_1 and h_2 independently run over all elements of H, which is just H itself, since H is a subgroup and is closed under composition. An expression such as $a(Hb)H$ is the set of elements $ah'h$, where h' runs over all elements of Hb and h independently runs over all elements of H, or equivalently, the set of elements $a(h_1b)h_2 = ah_1bh_2$, where h_1 and h_2 independently run over all elements of H. After overcoming our unfamiliarity with the notation, the point of what we are doing here can be appreciated. The composition of coset aH and coset bH, $(aH)(bH)$, is also a coset of H, the coset $(ab)H$ (although some element other than ab might be used to form it in the decomposition of G). Thus, the set of cosets of any invariant subgroup H of group G is closed under coset composition.

Note that we have risen to a higher level of complexity than that of groups with their elements and rules of composition of elements. Here we are looking at cosets as 'elements' of a set of cosets and are considering their composition. So our higher-level elements are cosets. They make up a set of cosets. Our higher-level composition operates between higher-level elements, i.e., between cosets.

We just found that the set of cosets of an invariant subgroup is closed under coset composition. Let us see if coset composition possesses any additional interesting properties. Is it associative? Yes:

$$[(aH)(bH)](cH) = (aH)[(bH)(cH)] .$$

as a direct result of associativity of the composition defined for G.

The composition of any coset of H with H itself gives

$$H(aH) = (Ha)H$$
$$= (aH)H$$
$$= a(HH)$$
$$= aH ,$$

and

$$(aH)H = a(HH) = aH ,$$

using associativity, equality of left and right cosets, and $HH = H$. So H has the characteristic property of an identity in coset composition; its composition with any coset in either order is just that coset.

And what happens when we compose any coset aH with coset $a^{-1}H$?

$$(aH)(a^{-1}H) = a(Ha^{-1})H$$
$$= a(a^{-1}H)H$$
$$= (aa^{-1})(HH)$$
$$= eH$$
$$= H$$

and

$$(a^{-1}H)(aH) = a^{-1}(Ha)H$$
$$= a^{-1}(aH)H$$
$$= (a^{-1}a)(HH)$$
$$= eH$$
$$= H ,$$

using associativity, equality of left and right cosets, properties of inverse and identity, and $HH = H$. Thus, for any coset aH there exists a coset $a^{-1}H$ (although some element other than a^{-1} might be used to form it in the decomposition of G), such that their coset composition in either way gives H. In other words, every coset has an inverse coset.

What have we found, if not a new group? This is a higher-level group than the group G that serves as the foundation of it all. This group is based on an invariant proper subgroup of G, on decomposition of G into cosets of the invariant subgroup, and on coset composition between cosets. In more detail, we start with any group G and invariant proper subgroup H. We decompose G into cosets of H (left and right cosets being the same),

$$G = H \cup aH \cup bH \cup \dots .$$

We consider the set consisting of H and its cosets, $\{H, aH, bH, \dots\}$. (Is it clear that each element of this set is itself a set, that this is a set of sets?) And we consider coset composition between pairs of cosets. And, indeed, we found (1) closure, (2) associativity, (3) existence of identity, and (4) existence of inverses – in short, a group. That group is called the *factor group* (also *quotient group*) of G by H and is denoted G/H. (I strongly suggest that at this point you review the development of the factor group and make sure it is clear how H's being a subgroup and the invariance of H each plays its role.) If G is of finite order n and H is of order m, the order of the factor group G/H is $s = n/m$ by Lagrange's theorem (see Sect. 9.3).

As an example, we take for G the non-Abelian order-6 group D_3, whose group table appears in Fig. 8.9 in Sect. 8.1 and Fig. 9.11 in Sect. 9.8, and its invariant subgroup $H = \{e, a, b\}$. The corresponding coset decomposition is

$$G = \{e, a, b\} \cup \{c, d, f\} .$$

The factor group G/H is the set of sets

$$G/H = \Big\{\{e, a, b\}, \ \{c, d, f\}\Big\}$$

under coset composition. It is an order-2 group, isomorphic with C_2.

9.4 Anatomy of Homomorphism

Let us return once more to homomorphism $G \to G'$ with kernel K. As was shown in Sect. 9.1, K is an invariant proper subgroup of G

(excluding the trivial homomorphism $G \to C_1$). So we decompose G into cosets of K,

$$G = K \cup aK \cup bK \cup \dots ,$$

and construct the factor group

$$G/K = \{K, aK, bK, \dots \} .$$

All elements of K are, by definition, mapped by the homomorphism to e', the identity element of G'. Consider the following structure-preservation diagram for any element a of G and all elements k of K, where a' is the image of a in G':

$$a \ k = (ak)$$
$$\downarrow \downarrow \quad \ \downarrow$$
$$a' \ e' = \ a'$$

Thus, all members of coset aK, which have the form ak with k running over all elements of K, are mapped to a'. Similarly, all members of coset bK are mapped to b', the image of b in G', and so on.

Can members of different cosets of K be mapped to the same element of G'? No, they cannot. To see that, assume b is not a member of coset aK, which is a necessary and sufficient condition for cosets aK and bK to be different, as we have seen. We then want to prove that, if $a \to a'$ and $b \to b'$, we must have $a' \neq b'$. So we assume the opposite, that a, $b \to a'$. By preservation of structure, the image of b^{-1} in G' must then be a'^{-1}. That gives us the structure preservation diagram for the composition $b^{-1}a$:

$$b^{-1} \ a = (b^{-1}a)$$
$$\downarrow \ \downarrow \quad \ \downarrow$$
$$a'^{-1} a' = \ e'$$

Since $b^{-1}a$ is mapped to e', it is an element of the kernel K, say $b^{-1}a = k$. Composition with b on the left and with k^{-1} on the right gives $ak^{-1} = b$. But if k is an element of K, so is k^{-1}, since K is a subgroup. Then b has the form ak_1, where k_1 is an element of K, and b is a member of coset aK in contradiction to our assumption that it is not.

Thus, whereas all members of the same coset of K are mapped by the homomorphism to the same element of G', members of different cosets have different images in G'. And since, by definition, all members of G' are images in the homomorphism, we thereby obtain a one-to-one correspondence between all cosets of K in G and all elements of G'. We also obtain the result that any homomorphism is an m-to-one mapping, where m is the order of the kernel as well as of every one of its cosets.

That one-to-one mapping between the factor group G/K and G' is, in fact, an isomorphism. Preservation of structure is based on the relation

$$(aK)(bK) = (ab)K \ ,$$

found in the preceding section, and is exhibited in the diagram

$$(aK) \ (bK) = (ab)K$$
$$\downarrow \quad \downarrow \quad \quad \downarrow$$
$$a' \quad b' \ = (a'b')$$

where a' and b' are the respective images of a and b in G'. The mapping looks like that of Fig. 9.1.

To summarize our result, given a homomorphism $G \to G'$ with kernel K, we have the isomorphism $G/K \sim G'$.

A converse result is obtained if, instead of starting with a homomorphism, we start with group G and an invariant proper subgroup of it, H. Decompose G into cosets of H,

$$G = H \cup aH \cup bH \cup \dots \ ,$$

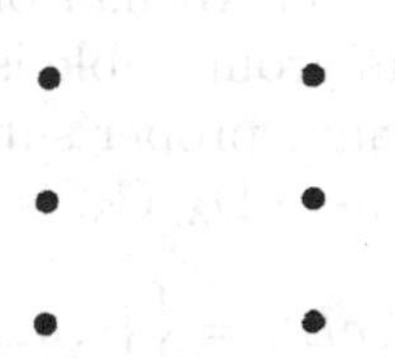

$$\frac{G/K} \sim \frac{G'}$$

$$K \longleftrightarrow e'$$

$$aK \longleftrightarrow a'$$

$$bK \longleftrightarrow b'$$

$$\bullet \qquad \bullet$$
$$\bullet \qquad \bullet$$
$$\bullet \qquad \bullet$$

Fig. 9.1. Isomorphism of G/K with G', where K is the kernel of the homomorphism of G to G'

and construct the factor group

$$G/H = \{H, aH, bH, \dots\} .$$

Consider the mapping from G to G/H, where each element of G is mapped to the coset of H of which it is a member. That is an m-to-one mapping, where m is the order of H (as well as of every coset of H), in which every element of G has an image in G/H and every element of G/H serves as an image. The mapping is a homomorphism. To confirm that structure is preserved, note that the cosets of H to which elements a, b, ab of G belong and to which they are mapped can be represented as aH, bH, $(ab)H$, respectively (although different elements might be used to form them). Then, by the relation

$$(aH)(bH) = (ab)H ,$$

which was demonstrated in Sect. 9.3, we have the structure preservation diagram

$$
\begin{array}{ccc}
a & b & = (ab) \\
\downarrow & \downarrow & \downarrow \\
(aH) & (bH) & = (ab)H
\end{array}
$$

Since structure is preserved, the mapping is indeed a homomorphism. The kernel of the homomorphism is H, since all elements of H, and only they, are mapped to H, the identity element of G/H.

To sum up this section, given a homomorphism $G \rightarrow G'$ with kernel K, we have the isomorphism $G/K \sim G'$. Conversely, given a group G and a invariant proper subgroup H, we have the homomorphism $G \rightarrow G/H$ with kernel H. That should help clarify what goes on in homomorphism and how homomorphism, kernel, invariant proper subgroup, and factor group are intimately interrelated.

Here are some examples to help clarify the clarification. See Sect. 8.1 for group tables.

1. We again return to the non-Abelian order-6 group D_3, discussed so often above, as G. (Its group table is shown also in Fig. 9.11 in Sect. 9.8.) Its only invariant proper subgroup is $H = \{e, a, b\}$, C_3. The factor group G/H, here D_3/C_3, is the order-2 group

$$G/H = \{H, cH\} = \Big\{\{e, a, b\}, \ \{c, d, f\}\Big\} ,$$

   isomorphic with the abstract order-2 group $G' = \{e', a'\}$, C_2. We then have the three-to-one homomorphism $G \rightarrow G/H$ with kernel

H, as well, of course, as the three-to-one homomorphism $G \rightarrow G'$ with kernel H. The whole situation can be represented as in Fig. 9.2.

2. Consider the order-4 group C_4, or its realization by the group of rotations about a common axis by $\{0°, 90°, 180°, 270°\}$, as G. Since the group is Abelian, its only proper subgroup is invariant. This subgroup is $H = \{e, b\}$, C_2, or its realization by the subgroup of rotations by $\{0°, 180°\}$. The factor group, C_4/C_2, is

$$G/H = \{H, aH\} = \big\{\{e, b\}, \{a, c\}\big\},$$

or its rotational realization

$$\big\{\{0°, 180°\}, \{90°, 270°\}\big\}.$$

It is isomorphic with the abstract group C_2, $G' = \{e', a'\}$, or its realization by rotations $\{0°, 180°\}$. We then have the two-to-one homomorphism $G \rightarrow G'$ with kernel H, among the abstract groups as well as among their rotational realizations. All that can be expressed by the diagram of Fig. 9.3.

3. We take an infinite-order example. The group of all nonzero real numbers under multiplication, R, includes the group of all positive numbers under multiplication, P, as a subgroup. It is an invariant

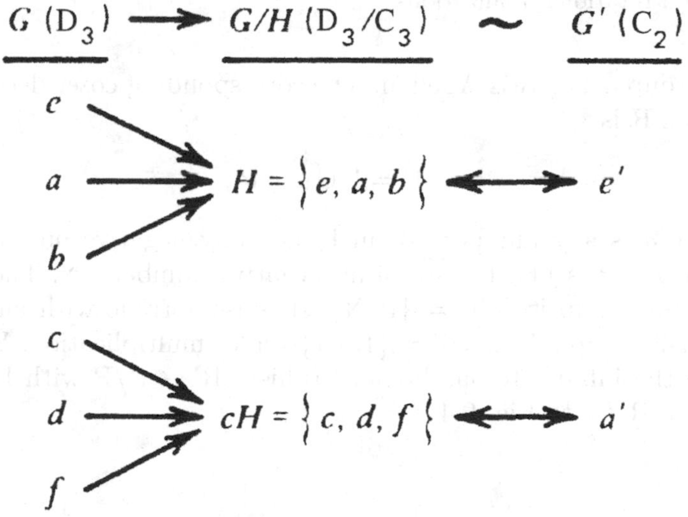

Fig. 9.2. Homomorphism of G (D$_3$) to G/H (D$_3$/C$_3$), isomorphism of G/H (D$_3$/C$_3$) with G' (C$_2$), and homomorphism of G (D$_3$) to G' (C$_2$)

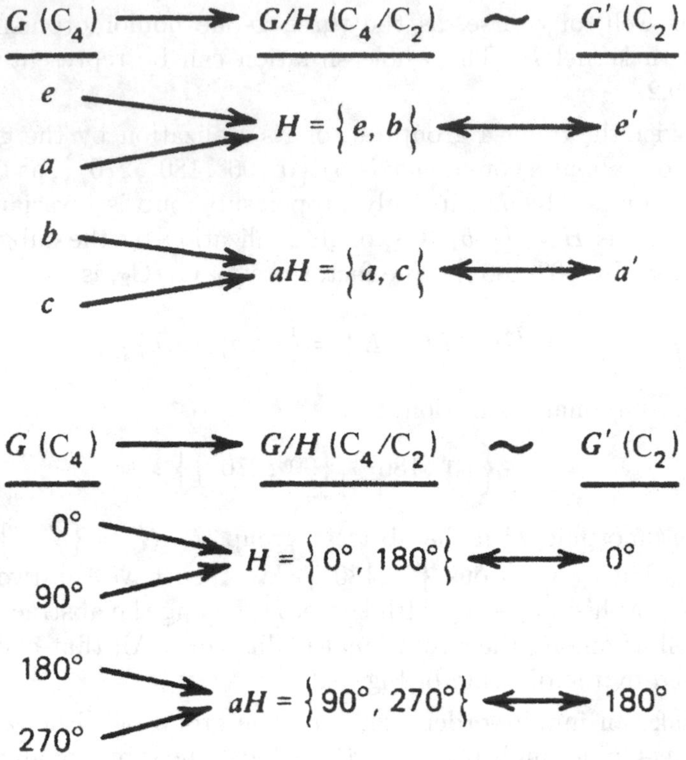

$$G\ (C_4) \longrightarrow G/H\ (C_4/C_2) \quad \sim \quad G'\ (C_2)$$

e
a
$\Longrightarrow H = \{e, b\} \longleftrightarrow e'$

b
c
$\Longrightarrow aH = \{a, c\} \longleftrightarrow a'$

$$G\ (C_4) \longrightarrow G/H\ (C_4/C_2) \quad \sim \quad G'\ (C_2)$$

0°
90°
$\Longrightarrow H = \{0°, 180°\} \longleftrightarrow 0°$

180°
270°
$\Longrightarrow aH = \{90°, 270°\} \longleftrightarrow 180°$

Fig. 9.3. Homomorphism of G (C_4) to G/H (C_4/C_2), isomorphism of G/H (C_4/C_2) with G' (C_2), and homomorphism of G (C_4) to G' (C_2): abstract groups and rotational realizations

subgroup, since R is Abelian. The corresponding coset decomposition of R is

$$R = P \cup a \times P,$$

where a is any number not in P, i.e., any negative number. The coset $a \times P$ is just the set of all negative numbers, N. The order-2 factor group is $R/P = \{P, N\}$. It is isomorphic with any other order-2 group, C_2, say $Z = \{1, -1\}$ under multiplication. We then have the infinity-to-one homomorphism $R \to R/P$ with P as the kernel. Refer to Fig. 9.4.

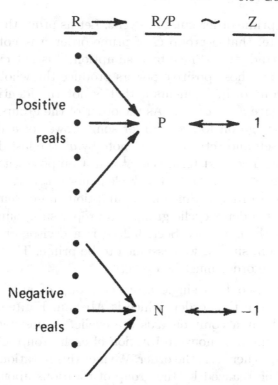

Fig. 9.4. Homomorphism of the multiplicative group of all nonzero real numbers, R, to R/P, where P is the multiplicative group of all positive numbers; isomorphism of R/P with the multiplicative group $Z = \{1, -1\}$; and homomorphism of R to Z. N is the set of all negative numbers

9.5 Generator

Any subset of elements of a finite-order group such that they, their positive powers, and their compositions (and repeated compositions, as many times as necessary) produce the whole group is called a *set of generators* of the group. For a given group a set of generators is not unique. For example, the set of all elements of a group is a set of generators, and so is the set of all elements excluding the identity. However, there are sometimes smaller sets, and there are always minimal sets. Minimal sets of generators are not in general unique either. If a minimal set of generators of a group consists of a single element, so that this element and its positive powers produce the whole group, such a group is called a *cyclic group*, and the element is called a *generating element* of the group. Since all elements of a cyclic group are powers of a single element, a cyclic group is always Abelian.

A group of prime order must be cyclic. Let us prove that by assuming the opposite, that a group G of prime order n is not cyclic, and showing a contradiction. Since by assumption G is not cyclic, it contains no element whose positive powers produce the whole the group. Consider any one of its elements a that is not the identity and consider the sequence $a, a^2, a^3, \ldots$. As the order of the group is finite, the sequence cannot go on forever. So for some power of a the sequence must repeat itself and return to a. Denote by a^k the last distinct term in the sequence. The next term is $a^{k+1} = a$. Compose with a^{-1} on the right to obtain $a^k = e$. Thus, the k elements $e, a, a^2, \ldots, a^{k-1}$ form a cyclic group of order k. Since by assumption those elements do not exhaust G, that order-k cyclic group is a proper subgroup of G. Then by Lagrange's theorem (see Sect. 9.2) k is a divisor of n. Here lies the contradiction, since n was assumed to be prime. That proves that a group of prime order must be cyclic.

Thus, there is only a single abstract group of any prime order, the cyclic group of that order, which is Abelian. Finite-order groups of rotations about a common axis are cyclic. A possible generating element is the smallest nonzero rotation of each group. Cyclic groups are denoted C_n, where n is the order. We met this notation in Sect. 8.1. Thus, C_n can be realized by the group of rotations about a common axis by

$$\left\{ 0°, 360°/n, 2 \times 360°/n, \ldots, (n-1) \times 360°/n \right\}.$$

(The notation D_n, which we have met also, is for the *dihedral groups*, of order $2n$. We do not discuss them here but will present their general definition in Sect. 10.5.)

Examples of minimal sets of generators and cyclic groups include the following. Relevant group tables are found in Sect. 8.1.

1. The order-2 group C_2 is cyclic, since $a^2 = e$, so a is its generating element. For its realization by $\{1, -1\}$ under multiplication, the number -1 is the generating element, since $(-1) \times (-1) = 1$. In terms of its realization by rotations about a common axis by $\{0°, 180°\}$, rotation by $180°$ is the generating element, since two consecutive rotations by $180°$ produce a rotation by $360°$, the identity transformation.

2. The order-3 group C_3 is cyclic, since $a^2 = b$ and $a^3 = e$, or $b^2 = a$ and $b^3 = e$, so that a and b are each a generating element. In terms of its realization by rotations by $\{0°, 120°, 240°\}$ about a common axis, rotation by $120°$ is a generating element, since two consecutive

rotations by $120°$ produce a rotation by $240°$, and three consecutive rotations by $120°$ result in a rotation by $360°$, which is rotation by $0°$. Rotation by $240°$ is also a generating element. Two consecutive rotations by $240°$ produce rotation by $480°$, which is rotation by $120°$, and three consecutive $240°$ rotations make a rotation by $720° = 2 \times 360°$, which is the identity transformation. Note that both 2 and 3 are primes and that there is only a single group of each order and it is cyclic.

3. The order-4 group C_4 is cyclic, since $a^2 = b$, $a^3 = c$, and $a^4 = e$, or $c^2 = b$, $c^3 = a$, and $c^4 = e$. Thus, a and c are generating elements. However, b is not, since $b^2 = e$ and higher powers of b do not produce a and c. In terms of the group's realization by rotations about a common axis by $\{0°, 90°, 180°, 270°\}$, the generating elements are rotation by $90°$ and rotation by $270°$. Rotation by $180°$ is not a generating element, since two consecutive rotations by $180°$ produce the identity transformation, and the other rotations are not produced by further consecutive rotations by $180°$.

4. Since four is not a prime, there may be noncyclic groups of that order, and indeed there is one, D_2. Minimal sets of generators are $\{a, b\}$ ($a^2 = e$, $ab = c$), $\{a, c\}$ ($a^2 = e$, $ac = b$), and $\{b, c\}$ ($b^2 = e$, $bc = a$). In terms of its realization by reflections and rotation, presented in Sect. 8.1 and in Figs. 8.6 and 8.7 there, a minimal set of generators consists of two reflections or the rotation and either reflection.

5. The order-5 group C_5 is of prime order and is cyclic. Any one of its nonidentity elements is a generating element, for example, c, $c^2 = a$, $c^3 = d$, $c^4 = b$, $c^5 = e$. In the group's rotational realization any nonzero rotation is a generating element.

6. The order-6 group D_3 is non-Abelian and therefore not cyclic. Its group table appears also in Fig. 9.11 in Sect. 9.8. Two of its minimal sets of generators are $\{a, c\}$ ($a^2 = b$, $a^3 = e$, $ac = f$, $a^2c = d$) and $\{d, f\}$ ($d^2 = e$, $df = a$, $fd = b$, $dfd = c$).

9.6 Direct Product

The *direct product* of two groups G and G' is denoted $G \times G'$ and is the group consisting of elements that are ordered pairs (g, g'), where g and g' run independently over all elements of G and G', respectively.

If the orders of G and G' are n and n', respectively, the order of $G \times G'$ is nn'. Composition for $G \times G'$ is defined in this way:

$$(a, a')(b, b') = (ab, a'b') \, ,$$

for all elements a, b in G and all a', b' in G'. Composition ab is in G, while composition $a'b'$ is in G'. The composition thus defined for the direct product group $G \times G'$ assures (1) closure and (2) associativity. (3) The identity of group $G \times G'$ is the element (e, e'), where e and e' are the respective identities of G and G'. (4) The inverse of any element (g, g') of $G \times G'$ is (g^{-1}, g'^{-1}). Thus, $G \times G'$ is shown to be a group, as claimed.

Among the possible proper subgroups of $G \times G'$ there is one iso-morphic with G, consisting of elements of the form (g, e') for all g in G, and another isomorphic with G', consisting of elements of the form (e, g') for all g' in G'. Both are invariant subgroups. The direct-product group $G' \times G$ is isomorphic with $G \times G'$. [To see that, consider the mapping $(g', g) \leftrightarrow (g, g')$ and confirm that it preserves structure.]

Higher-degree direct-product groups are formed similarly. For in-stance, $G \times G' \times G''$ denotes the group of ordered triples (g, g', g''), where g, g', g'' run independently over all elements of G, G', G'', re-spectively. Composition for that group is

$$(a, a', a'')(b, b', b'') = (ab, a'b', a''b'') \, .$$

Consider the following examples:

1. The direct product $C_2 \times C_2$ is an order-4 group. But which one of the two abstract order-4 groups is it? Refer to Sect. 8.1. The elements of $C_2 \times C_2$ are (e, e), (e, a), (a, e), and (a, a), with $a^2 = e$. The group table is shown in Fig. 9.5. The total domination of the group-table diagonal by the identity element (e, e) discloses that this is the structure of D_2 (Fig. 8.5 in Sect. 8.1). One possible isomorphism mapping is shown in Fig. 9.6.

(e, e)	(e, a)	(a, e)	(a, a)
(e, a)	(e, e)	(a, a)	(a, e)
(a, e)	(a, a)	(e, e)	(e, a)
(a, a)	(a, e)	(e, a)	(e, e)

Fig. 9.5. Group table of $C_2 \times C_2$

$$C_2 \times C_2 = \Big\{ (e,\, e),\ (e,\, a),\ (a,\, e),\ (a,\, a) \Big\}$$

$$\wr \qquad \updownarrow \qquad \updownarrow \qquad \updownarrow \qquad \updownarrow$$

$$D_2 \;=\; \Big\{\ e',\qquad a',\qquad b',\qquad c'\ \Big\}$$

Fig. 9.6. Isomorphism of $C_2 \times C_2$ with D_2

(e, e')	(a, a')	(e, b')	(a, e')	(e, a')	(a, b')
(a, a')	(e, b')	(a, e')	(e, a')	(a, b')	(e, e')
(e, b')	(a, e')	(e, a')	(a, b')	(e, e')	(a, a')
(a, e')	(e, a')	(a, b')	(e, e')	(a, a')	(e, b')
(e, a')	(a, b')	(e, e')	(a, a')	(e, b')	(a, e')
(a, b')	(e, e')	(a, a')	(e, b')	(a, e')	(e, a')

Fig. 9.7. Group table of $C_2 \times C_3$

e''	a''	b''	c''	d''	f''
a''	b''	c''	d''	f''	e''
b''	c''	d''	f''	e''	a''
c''	d''	f''	e''	a''	b''
d''	f''	e''	a''	b''	c''
f''	e''	a''	b''	c''	d''

Fig. 9.8. Group table of C_6

2. The direct product $C_2 \times C_3$ is an order-6 group. Since the constituent groups are Abelian, so is their direct product. Therefore, that is not the order-6 group D_3, discussed so frequently already (see Sect. 8.1). The elements of $C_2 \times C_3$ are (e, e'), (a, a'), (e, b'), (a, e'), (e, a'), (a, b'), where $\{e, a\}$ is C_2 and $\{e', a', b'\}$ is C_3. The group table is shown in Fig. 9.7. If we relabel the elements e'', a'', b'', c'', d'', f'', respectively, the group table of Fig. 9.7 takes the form of Fig. 9.8. That is clearly the cyclic group of order 6, C_6 (see Sect. 9.5). Its generating elements are a'' and

f'', or in $C_2 \times C_3$ notation (a, a') and (a, b'). The group C_6 can be realized by the group of rotations about a common axis by $\{0°, 60°, 120°, 180°, 240°, 300°\}$, whose generating elements are rotation by $60°$ and rotation by $300°$. The groups C_6 and D_3 (see Fig. 8.9 in Sect. 8.1) are the only abstract groups of order 6.

3. The group of complex numbers under addition is isomorphic with the direct product of the group of real numbers under addition with itself.

4. The group of 3-vectors under vector addition is isomorphic with the direct product of the group of real numbers under addition with itself twice.

9.7 Permutation, Symmetric Group

A *permutation* is a rearrangement of objects. Imagine, for example, that we have four objects in positions 1, 2, 3, 4, and further imagine that we permute, i.e., rearrange, them so that the object in position 1 is placed in position 4, the object in position 2, is moved to position 1, the object in position 3 is left where it is, and the object in position 4 finds itself in position 2. Figure 9.9 should help here. Such a permutation is denoted $\begin{pmatrix} 1234 \\ 4132 \end{pmatrix}$, where the top row lists the positions and each number in the bottom row is the final location of the object originally located at the position whose number is directly above it in the top row. The top row can be in any order, so both $\begin{pmatrix} 2431 \\ 1234 \end{pmatrix}$ and $\begin{pmatrix} 3142 \\ 3421 \end{pmatrix}$, for instance, denote the same permutation as in our example.

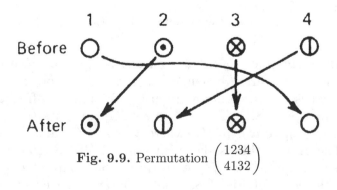

Fig. 9.9. Permutation $\begin{pmatrix} 1234 \\ 4132 \end{pmatrix}$

A general permutation of n objects is denoted

$$\begin{pmatrix} 1 & 2 & 3 & \dots & n \\ p_1 & p_2 & p_3 & \dots & p_n \end{pmatrix},$$

where p_1, p_2, $\dots$, p_n are the numbers $1, 2, \dots, n$ in some order (i.e., some permutation of $1, 2, \dots, n$), which directs us to move the object in position 1 to position p_1, the object in position 2 to position p_2, and so on. The top row can be in any order, as long as p_i appears beneath i for $i = 1, 2, \dots, n$.

Permutations of the same number of objects are composed by consecutive application; the second permutation acts on the result of the first. The composition of permutations

$$a = \begin{pmatrix} 1234 \\ 4132 \end{pmatrix} \quad \text{and} \quad b = \begin{pmatrix} 1234 \\ 1342 \end{pmatrix},$$

for example, with permutation a acting first and permutation b acting on the result of a, is denoted

$$ba = \begin{pmatrix} 1234 \\ 1342 \end{pmatrix} \begin{pmatrix} 1234 \\ 4132 \end{pmatrix}.$$

(The order of application of permutations, as well as of any other kind of transformation, is specified from right to left. That will be discussed in Sect. 10.2.) The composition is the permutation obtained by following the itinerary of each object through the permutations to which it is subjected. In the example the object originally in position 1 is carried to position 4 by permutation a and from there to position 2 by b. The object originally in position 2 is shifted to position 1 by a and then left in position 1 by b. And so on to obtain

$$ba = \begin{pmatrix} 1234 \\ 2143 \end{pmatrix}.$$

A useful trick is to reorder the rows of the second permutation to make the top row match the bottom row of the first permutation. The combined permutation is then obtained by 'canceling' the bottom row of the first permutation with the top row of the second, leaving the top row of the first and the bottom row of the second as the result. In the example,

$$ba = \begin{pmatrix} 1234 \\ 1342 \end{pmatrix} \begin{pmatrix} 1234 \\ 4132 \end{pmatrix}$$

$$= \begin{pmatrix} 4132 \\ 2143 \end{pmatrix} \begin{pmatrix} 1234 \\ 4132 \end{pmatrix}$$

$$= \begin{pmatrix} 1234 \\ 2143 \end{pmatrix}.$$

And as another example,

$$ab = \begin{pmatrix} 1234 \\ 4132 \end{pmatrix} \begin{pmatrix} 1234 \\ 1342 \end{pmatrix}$$

$$= \begin{pmatrix} 1342 \\ 4321 \end{pmatrix} \begin{pmatrix} 1234 \\ 1342 \end{pmatrix}$$

$$= \begin{pmatrix} 1234 \\ 4321 \end{pmatrix}.$$

For general permutations of n objects

$$a = \begin{pmatrix} 1 & 2 & 3 & \dots & n \\ p_1 & p_2 & p_3 & \dots & p_n \end{pmatrix} \quad \text{and} \quad b = \begin{pmatrix} 1 & 2 & 3 & \dots & n \\ q_1 & q_2 & q_3 & \dots & q_n \end{pmatrix},$$

the composition is

$$ba = \begin{pmatrix} 1 & 2 & 3 & \dots & n \\ q_1 & q_2 & q_3 & \dots & q_n \end{pmatrix} \begin{pmatrix} 1 & 2 & 3 & \dots & n \\ p_1 & p_2 & p_3 & \dots & p_n \end{pmatrix}$$

$$= \begin{pmatrix} p_1 & p_2 & p_3 & \dots & p_n \\ q_{p_1} & q_{p_2} & q_{p_3} & \dots & q_{p_n} \end{pmatrix} \begin{pmatrix} 1 & 2 & 3 & \dots & n \\ p_1 & p_2 & p_3 & \dots & p_n \end{pmatrix}$$

$$= \begin{pmatrix} 1 & 2 & 3 & \dots & n \\ q_{p_1} & q_{p_2} & q_{p_3} & \dots & q_{p_n} \end{pmatrix}.$$

Similarly,

$$ab = \begin{pmatrix} 1 & 2 & 3 & \dots & n \\ p_{q_1} & p_{q_2} & p_{q_3} & \dots & p_{q_n} \end{pmatrix}.$$

Thus, the composition of permutations of n objects is also a permutation of n objects. Composition of permutations is not, in general, commutative, as we see in the example or in the general case. It is,

however, associative, as we could easily prove by composing three permutations and obtaining three tiers of indices in the bottom row, which
we mercifully refrain from doing. (The composition of transformations
of any kind by consecutive application is associative, as will be seen in
Sect. 10.2.)

The 'nonpermutation' $\begin{pmatrix} 1 \ 2 \ \dots \ n \\ 1 \ 2 \ \dots \ n \end{pmatrix}$ serves as the identity permutation of n objects. The inverse of the permutation

$$\begin{pmatrix} 1 & 2 & \dots & n \\ p_1 & p_2 & \dots & p_n \end{pmatrix}$$

is simply the permutation

$$\begin{pmatrix} p_1 & p_2 & \dots & p_n \\ 1 & 2 & \dots & n \end{pmatrix}.$$

Each clearly undoes the other.

What could we be leading up to, if not that the set of permutations of n objects, under composition of consecutive permutation,
forms a group? It is called the *symmetric group* of degree n and is
denoted S_n. In the preceding paragraphs we (1) proved closure, (2)
claimed associativity, and showed (3) existence of identity and (4) existence of inverses. The order of S_n is $n!$. It is non-Abelian for $n > 2$.
The symmetric group S_n includes S_m as a subgroup, $S_n \supset S_m$, for
$n \geq m$.

Here are several examples of symmetric groups:

1. S_1 is the order-1 group consisting of the identity permutation of
   a single object, $\begin{pmatrix} 1 \\ 1 \end{pmatrix}$.

2. S_2 consists of the two permutations $\begin{pmatrix} 1 \ 2 \\ 1 \ 2 \end{pmatrix}$ and $\begin{pmatrix} 1 \ 2 \\ 2 \ 1 \end{pmatrix}$ and is isomorphic with C_2. It is Abelian.

3. S_3 is an order-6 group ($3! = 6$). It is non-Abelian, so it must be isomorphic with D_3, whose group table appears in Fig. 8.9 in Sect. 8.1
   and in Fig. 9.11 in Sect. 9.8. An isomorphism mapping is shown in
   Fig. 9.10.

4. S_4 is of order 24 ($= 4!$), too large to consider here, and higher-
   degree symmetric groups possess even larger, and rapidly increasing, orders. For example, $11! = 39\,916\,800$, and $12!$ exceeds the
   capacity of an eight-digit calculator.

$$S_3 = \left\{ \begin{pmatrix} 123 \\ 123 \end{pmatrix}, \begin{pmatrix} 123 \\ 231 \end{pmatrix}, \begin{pmatrix} 123 \\ 312 \end{pmatrix}, \begin{pmatrix} 123 \\ 132 \end{pmatrix}, \begin{pmatrix} 123 \\ 321 \end{pmatrix}, \begin{pmatrix} 123 \\ 213 \end{pmatrix} \right\}$$

$$\wr \qquad \updownarrow \qquad \updownarrow \qquad \updownarrow \qquad \updownarrow \qquad \updownarrow \qquad \updownarrow$$

$$D_3 = \{ \quad e, \qquad a, \qquad b, \qquad c, \qquad d, \qquad f, \ \}$$

Fig. 9.10. Isomorphism of S_3 with D_3

9.8 Cayley's Theorem

Consider the group table of any group G of finite order n. For example, the group D_3, whose table is repeated in Fig. 9.11. If the first row is $e, a, b, \ldots$, the row starting with element g is obtained from the first row by composition with g on the left: $ge \ (= g)$, ga, gb, ... It follows directly that (1) each row contains all the elements of the first row, but (2) in some other arrangement (unless $g = e$), and (3) rows starting with different elements are different. As for (1), if any row did not contain all the elements of the first row, we would have, say, $ga = gb$ for the element g at the left of that row and some pair of elements a, b. Composing both sides of the equation with g^{-1} on the left, we would then have $a = b$, which is a contradiction, since the first row of the group table contains n different symbols for n distinct elements. (2) The arrangement of the elements in the row starting with g is different from that of the first row, simply because the first row starts with e and the other does not (unless $g = e$, and we are talking about the first row anyway). Result (3) is trivial, since rows starting with different elements differ simply in that they start with different elements. Confirm (1), (2), and (3) in Fig. 9.11 for D_3.

e	a	b	c	d	f
a	b	e	f	c	d
b	e	a	d	f	c
c	d	f	e	a	b
d	f	c	b	e	a
f	c	d	a	b	e

Fig. 9.11. Group table of D_3

What we are getting at, however, is that to each element of a finite group we can assign a unique permutation of n objects in a one-to-one correspondence; to every element g we assign the permutation whose top row is the top row of the group table and whose bottom row is the row of the group table that starts with g,

$$g \longleftrightarrow \begin{pmatrix} e & a & b & \dots \\ g & ga & gb & \dots \end{pmatrix}.$$

That is indeed a permutation, although we have become used to denoting permutations with numbers rather than with letters. To make it look more familiar one might replace the letters $e, a, b, \dots$ by the numbers $1, 2, \dots, n$ according to some cipher.

From the discussion of the second preceding paragraph it is clear that this is indeed a one-to-one mapping of all elements of G onto a subset of S_n. In fact, the mapping is an isomorphism, and the subset of S_n is therefore a subgroup of S_n. To see that structure is preserved, consider Fig. 9.12, keeping in mind our trick for composing permutations. (The whole discussion could be carried out just as well with columns instead of rows of a group table.) The result, known as Cayley's theorem (Arthur Cayley, British mathematician, 1821–1895), is that every group of finite order n is isomorphic with a subgroup of the symmetric group S_n. Cayley's theorem is helpful for finding groups of a given order.

$$
\begin{array}{c}
g_2 \\ \updownarrow
\end{array}
\quad
\begin{array}{c}
g_1 \\ \updownarrow
\end{array}
\quad = \quad
\begin{array}{c}
g_2 \\ \updownarrow
\end{array}
\quad
\begin{array}{c}
g_1 \\ \updownarrow
\end{array}
$$

$$
\begin{pmatrix} e & a & b & \cdots \\ g_2 & g_2 a & g_2 b & \cdots \end{pmatrix}
\begin{pmatrix} e & a & b & \cdots \\ g_1 & g_1 a & g_1 b & \cdots \end{pmatrix}
=
\begin{pmatrix} g_1 & g_1 a & g_1 b & \cdots \\ g_2 g_1 & g_2(g_1 a) & g_2(g_1 b) & \cdots \end{pmatrix}
\begin{pmatrix} e & a & b & \cdots \\ g_1 & g_1 a & g_1 b & \cdots \end{pmatrix}
$$

$$
= \quad (g_2 g_1) \\ \updownarrow
$$

$$
= \begin{pmatrix} e & a & b & \cdots \\ g_2 g_1 & (g_2 g_1)a & (g_2 g_1)b & \cdots \end{pmatrix}
$$

Fig. 9.12. Structure preservation diagram for Cayley's theorem

9.9 Summary

In this chapter we continued the introduction to group theory that we started in the preceding chapter. We summarize by listing the more important concepts that are discussed in each section:

- Section 9.1: conjugate elements, conjugacy class, equivalence class, invariant (normal) subgroup, kernel of homomorphism.
- Section 9.2: coset, coset decomposition, Lagrange's theorem.
- Section 9.3: coset composition, factor (quotient) group.
- Section 9.4: homomorphism.
- Section 9.5: set of generators, cyclic group, generating element.
- Section 9.6: direct product.
- Section 9.7: permutation, symmetric group.
- Section 9.8: Cayley's theorem.

The Formalism of Symmetry

Now that we have laid the mathematical foundation for the formal study of symmetry, we are prepared to deal with a general symmetry formalism. Such a formalism is needed for many, especially quantitative, applications of symmetry considerations in science. In this chapter we will develop a general symmetry formalism, making extensive use of the mathematical concepts that were discussed in Chaps. 8 and 9 and earlier in the book. Our road to a symmetry formalism starts with a state space of a system and proceeds through transformation, transformation group, symmetry transformation, symmetry group, and approximate symmetry transformation. The quantification of symmetry will be discussed, as will the application of the symmetry formalism to quantum systems.

10.1 System, State

We start by presenting the concepts of system, subsystem, state, and state space. Those concepts are purposely not sharply defined and are assumed to be more or less intuitively understood, in order to allow the widest possible applicability of the following development.

A *system* is whatever we investigate the properties of. I am intentionally being vague and general here, so as to impose no limitations on the possible objects of our interest. A system might be abstract or concrete, microscopic or macroscopic, static or dynamic, finite or infinite. Anything – and not only things – can be a system.

A *subsystem* is a system that is wholly contained, in some sense, within a system. Again, vagueness and generality.

A *state* of a system is a possible condition of the system. Here, too, we are saying very little. We can add, though, that a system might possess a finite or infinite number of states, and that the same system might have different kinds or numbers of states, depending on how it is being considered.

A *state space* of a system is a set of all states of the same kind, where 'the same kind' may be interpreted in any useful way. Earlier in this book we used the term 'set of states' for what we now call state space.

Consider some examples:

1. Let the system be a given amount of a certain pure gas. Considered microscopically, the states of the system are characterized by the positions and velocities of all the molecules (assuming spinless, structureless, point particles). The set of all allowable positions and velocities for all molecules forms a state space of the system. Macrostates, assuming that the gas is confined and in equilibrium, are described by any two quantities among, say, pressure, volume, and temperature. So a state space might be the set of all allowed pressures and temperatures. In both cases, microscopic and macroscopic, the number of states is infinite, though 'more so' in the microscopic case.

2. Take as the system a plane figure. States of such a system are described by shape, size, location, orientation, and even color, texture, and so on, and are infinite in number.

3. Or to be more explicit, let the system be a plane figure of given shape and size, lying in a given plane, and with one of its points fixed in the plane. Ignore color, texture, and the like. States of this system are characterized by a single angle specifying its rotational orientation in the plane and by its *handedness* (also called *chirality*), i.e., which of its two mirror-twin versions (such as left and right hands) is appearing.

4. Consider the system of a ball lying in any one of three depressions in the sand. That system possesses three states.

5. Consider the system of three balls lying in three depressions. States are described by specifying which ball is in which depression. The state space of that system consists of six states, the six possible arrangements of three balls.

6. The system might be any quantum system. Its states are quantum states, characterized by a set of quantum numbers, the eigenvalues

of a complete set of commuting Hermitian operators. State space is postulated to be a Hilbert space, in which states are represented by vectors (actually rays). Quantum systems will be treated in more detail in Sect. 10.8.

10.2 Transformation, Transformation Group

See Sect. 8.2 for our earlier discussion of mapping. Here we will make free use of concepts and terminology that were presented and discussed there.

A *transformation* of a system is a mapping of a state space of the system into itself. What that means is this: Every state has an image, which is also a state; more than one state might have the same image; but not every state must necessarily serve as an image in the mapping. We denote the action of a transformation T by

$$u \xrightarrow{\ T\ } v \quad \text{or} \quad v = T(u) \,,$$

in arrow notation and function notation, where u and v represent states of the system. State v is the image of state u under transformation T. State u runs over all states of a state space. If the number of states of a state space is finite, a transformation might be expressed as a two-column table. Otherwise it must be expressed as a general rule.

Transformations are composed by consecutive application, i.e., the composition of two transformations is defined as the result of applying one transformation to the result of the other. For transformations T_1 and T_2 one possible composition is $T_2 T_1$, denoting the result of first applying T_1 and then applying T_2, as shown in Fig. 10.1. The function notation practically 'forces' us to specify consecutively applied transformations from right to left, and we adhere to that convention. The

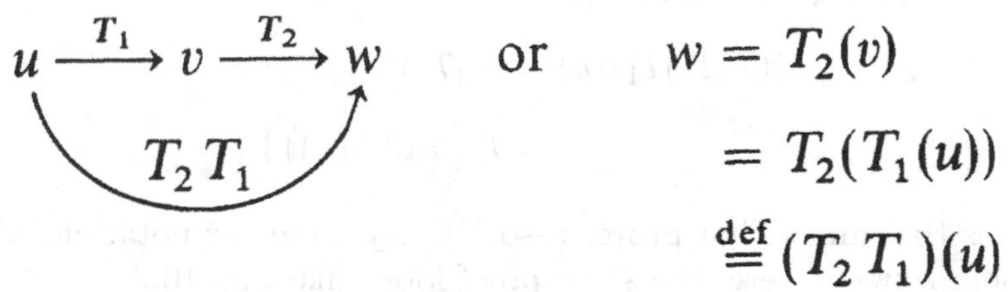

$$u \xrightarrow{\ T_1\ } v \xrightarrow{\ T_2\ } w \qquad \text{or} \qquad w = T_2(v)$$
$$\underbrace{\qquad\qquad}_{T_2 T_1} \qquad\qquad\qquad = T_2(T_1(u))$$
$$\overset{\text{def}}{=} (T_2 T_1)(u)$$

Fig. 10.1. Shown in arrow notation, the transformation $T_2 T_1$ applied to an arbitrary state u is the result of first applying transformation T_1 to u and then applying transformation T_2 to image state v to obtain image state w

other possible composition of T_1 and T_2 is, of course, T_1T_2, where T_2 is applied first, followed by T_1.

For composition of transformations to be meaningful, they must act in the same state space. That will always be tacitly assumed. As defined, the composition of any two transformations is a mapping of a state space into itself and is therefore also a transformation. Thus, the set of transformations of a system for any of its state spaces is closed under the composition of consecutive application.

In general it matters in which order transformations are applied,

$$T_2T_1 \neq T_1T_2 .$$

However, a pair of transformations T_1, T_2 whose composition in either order gives the same result $T_2T_1 = T_1T_2$, meaning

$$(T_2T_1)(u) = (T_1T_2)(u) ,$$

for all states u of a state space, are said to *commute*.

Composition of transformations by consecutive application is easily shown to be associative, i.e.,

$$T_3(T_2T_1) = (T_3T_2)T_1 ,$$

for all transformations T_1, T_2, T_3, so that the notation $T_3T_2T_1$ is unambiguous. We referred to and made use of that property of transformation composition a number of times earlier in this book. Using the definition of composition, the transformation on the left-hand side of the equation acting on arbitrary state u is

$$\big(T_3(T_2T_1)\big)(u) = T_3\big((T_2T_1)(u)\big)$$
$$= T_3\Big(T_2\big(T_1(u)\big)\Big) .$$

Similarly, the right-hand side gives

$$\big((T_3T_2)T_1\big)(u) = (T_3T_2)\big(T_1(u)\big)$$
$$= T_3\Big(T_2\big(T_1(u)\big)\Big) ,$$

which is the same. That proves associativity. In arrow notation, which is more awkward here, the same proof looks like Fig. 10.2.

The mapping I of every state of a state space to itself,

$$u \xrightarrow{\ I\ } u \quad \text{or} \quad u = I(u) ,$$

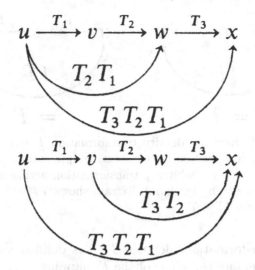

Fig. 10.2. Proof of associativity of transformation composition by consecutive application, shown in arrow notation. Both diagrams are valid for arbitrary transformations T_1, T_2, T_3 and any state u. States v, w, x are images and images of images. The top diagram shows $T_3(T_2T_1) = T_3T_2T_1$ and the bottom diagram shows $(T_3T_2)T_1 = T_3T_2T_1$

for all states u of a state space, is called the *identity transformation*. Indeed, it acts as the identity under composition by consecutive application:

$$TI = IT = T \ ,$$

for all transformations T, since for arbitrary state u

$$(TI)(u) = T\big(I(u)\big) = T(u) \ , \qquad (IT)(u) = I\big(T(u)\big) = T(u) \ .$$

That is shown in arrow notation in Fig. 10.3.

Among the transformations of a system there are those that are one-to-one and onto (bijective), i.e., every state of a state space serves as an image of the mapping and is the image of only a single state. For those transformations, and only for those, can inverses be defined. For any transformation T that is one-to-one and onto,

$$u \xrightarrow{\;T\;} v \quad \text{or} \quad v = T(u) \ ,$$

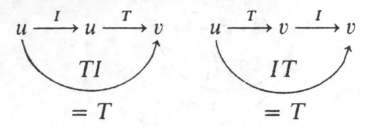

Fig. 10.3. Proof that the identity transformation I acts as the identity of transformation composition by consecutive application, shown in arrow notation. T represents an arbitrary transformation acting on any state u, giving image state v. The left-hand diagram shows $TI = T$ and the right-hand diagram shows $IT = T$

the inverse transformation, denoted T^{-1}, is defined as the mapping obtained by reversing the sense of the T mapping,

$$u \xrightarrow{T^{-1}} v \quad \text{or} \quad u = T^{-1}(v) \, .$$

That was explained in detail in Sect. 8.2.

Under composition by consecutive application, transformations T and T^{-1} are mutual inverses, since their compositions in both orders result in the identity transformation,

$$T^{-1}T = TT^{-1} = I \, .$$

That is proved in Figs. 10.4 and 10.5.

I hope it is clear that we are discovering a group. The set of all invertible transformations of any state space of a system forms a group under composition of consecutive application, called a *transformation group* of the system.

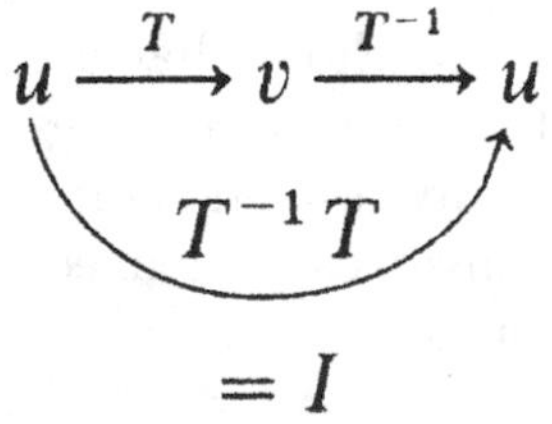

$$u \xrightarrow{T} v \xrightarrow{T^{-1}} u \qquad \text{or} \qquad (T^{-1}T)(u) = T^{-1}(T(u))$$
$$\underset{T^{-1}T}{\underbrace{}} \qquad\qquad = T^{-1}(v)$$
$$= I \qquad\qquad\qquad\qquad = u$$
$$= I(u)$$

Fig. 10.4. Proof that $T^{-1}T = I$ for any invertible transformation T, where I is the identity transformation, u is any state, and v is the image of u by T

$$v \xrightarrow{T^{-1}} u \xrightarrow{T} v \qquad \text{or} \qquad (TT^{-1})(v) = T(T^{-1}(v))$$

$$\underbrace{\qquad\qquad}_{TT^{-1}} \qquad\qquad\qquad = T(u)$$

$$= I \qquad\qquad\qquad\qquad\qquad = v$$

$$= I(v)$$

Fig. 10.5. Proof that $TT^{-1} = I$ for any invertible transformation T, where I is the identity transformation, v is any state, and u is the image of v by T^{-1}

1. The set is closed under composition, since the composition of invertible transformations is an invertible transformation. Indeed, the composition of one-to-one (injective) transformations is a one-to-one transformation, and, if every state of a state space serves as an image for each of two transformations, every state will serve as an image for their composition. [Or in short, the composition of onto (surjective) transformations is an onto transformation.]
2. We proved associativity.
3. The identity transformation is invertible, so belongs to the set, and serves as the identity element.
4. Existence of inverses holds by definition, since we are dealing with the set of all *invertible* transformations.

Note that a transformation group of a system is not unique, i.e., *the* transformation group of a system is not defined in general, as the same system might possess different transformation groups for different state spaces.

I might point out that a transformation that is one-to-one and onto (bijective), i.e., invertible, is a permutation of state space. If such a transformation, acting on a finite state space, is represented by a two-column table, the list of states in the image column will be some permutation of the list of states in the object column. That is true, since for such a transformation all states appear in each column and, moreover, appear only once in each. Thus the transformation group of any finite state space containing n states is isomorphic with the symmetric group of degree n, S_n (which we discussed in Sect. 9.7 and is not to be confused with symmetry group, which we will study in Sect. 10.5).

For infinite state spaces the idea is the same, although perhaps not so clear intuitively.

Examples of transformation groups include the following:

1. Consider the equilibrium macrostates of the system of a given amount of a certain confined pure gas. Since p, V, T (pressure, volume, absolute temperature) are related by an equation of state, we are free to vary any two of them, as long as they remain positive. Thus, all transformations of the system are of the form

$$p \longrightarrow p' = f(p, T) , \qquad T \longrightarrow T' = g(p, T) ,$$

for arbitrary positive functions $f(p, T)$ and $g(p, T)$. For all invertible positive functions whose range of values is all the positive numbers we obtain the (infinite-order) transformation group of the system for its macrostate space.

2. Now considering microstates of the same system, each molecule (all of which are identical, since the gas is pure) has three degrees of freedom (assuming spinless, structureless, point particles), so its state is characterized by three coordinates, x, y, z, and three components of velocity, v_x, v_y, v_z. All coordinates and all velocity components for all molecules can be varied freely (ignoring the relativistic limit of the speed of light). Thus, all transformations of the system have the form

$$x_i \longrightarrow x_i' = X_i(x_j, y_j, z_j, v_{xj}, v_{yj}, v_{zj}) ,$$
$$y_i \longrightarrow y_i' = Y_i(x_j, y_j, z_j, v_{xj}, v_{yj}, v_{zj}) ,$$
$$z_i \longrightarrow z_i' = Z_i(x_j, y_j, z_j, v_{xj}, v_{yj}, v_{zj}) ,$$
$$v_{xi} \longrightarrow v_{xi}' = V_{xi}(x_j, y_j, z_j, v_{xj}, v_{yj}, v_{zj}) ,$$
$$v_{yi} \longrightarrow v_{yi}' = V_{yi}(x_j, y_j, z_j, v_{xj}, v_{yj}, v_{zj}) ,$$
$$v_{zi} \longrightarrow v_{zi}' = V_{zi}(x_j, y_j, z_j, v_{xj}, v_{yj}, v_{zj}) ,$$

for $i, j = 1, \ldots, N$, where N is the number of molecules, with arbitrary functions $X_i(\ )$, $Y_i(\ )$, $Z_i(\ )$, $V_{xi}(\ )$, $V_{yi}(\ )$, $V_{zi}(\ )$. For all invertible functions whose range of values is all the real numbers we obtain the (infinite-order) transformation group of the system for its microstate space.

3. For a plane figure of given shape and size, lying in a given plane, with one of its points fixed in the plane, the angle of rotational orientation in the plane, γ, and its handedness can be varied freely. Thus, all transformations of this system consist of the angle changes

$$\gamma \longrightarrow \gamma' = g(\gamma) \ .$$

[these are orientation dependent rotations within the given plane about the fixed point by angle $g(\gamma) - \gamma$] and reflections through all two-sided mirrors perpendicular to the given plane and passing through the fixed point. Restricting the functions $g(\gamma)$ to being invertible, we obtain, including the reflections, a transformation group (of infinite order) of the system.

If, however, the transformation

$$\gamma \longrightarrow \gamma' = g(\gamma)$$

is understood, not as giving the new orientation of the whole figure for every possible original orientation, but rather as giving the new angular position (with respect to the fixed point) of each point of the figure for every possible original angular position of the point, whatever the state of the figure as a whole – which, in fact, is the usual interpretation (see Sect. 10.3) – then for the figure to retain its shape, all its points must rotate through the same angle. The functions $g(\gamma)$ are then immediately restricted to the form

$$g(\gamma) = \gamma + \alpha \ ,$$

which describes rotation by angle α,

$$\gamma \longrightarrow \gamma' = \gamma + \alpha \ .$$

4. For the system of one ball and three depressions the transformation group is the group of permutations of the three states of its state space, the symmetric group of degree 3, S_3, of order $3! = 6$ (see Sect. 9.7).

5. For the system of three balls in three depressions, the transformation group is the group of permutations of the six states of its state space, which is the symmetric group of degree 6, S_6, whose order is $6! = 720$ (see Sect. 9.7).

6. For any quantum system the set of all transformations is all the operators in its Hilbert space of states. The subset of all invertible operators constitutes the transformation group. However, general operators are not considered for quantum systems, but only linear and antilinear ones. The transformation group is correspondingly the subset of all invertible linear and antilinear operators. See Sect. 10.8 for a more detailed treatment of quantum systems.

10.3 Transformations in Space, Time, and Space-Time

Although we have defined transformations in a very general way, and, as we saw in the examples in Sect. 10.2, transformations can be very diverse indeed, transformations in space, in time, and in space-time are especially important, since the states of many systems are described in spatiotemporal terms. We therefore devote a section to a brief summary of some of the more important and useful transformations of these kinds. We use Cartesian coordinates (x, y, z) (after René Descartes, French philosopher, mathematician, and scientist, 1596–1650) for space and Minkowskian coordinates (x, y, z, t) (Hermann Minkowski, Russian mathematician, 1864–1909) for space-time. First consider a number of spatial transformations.

The transformation of *spatial displacement* (or *spatial translation*) maps all points of space to image points that are all the same distance away and in the same direction from their object points. Spatial displacements are invertible transformations. The most general spatial displacement has the form

$$x \longrightarrow x' = x + a , \qquad x' \longrightarrow x = x' - a ,$$
$$y \longrightarrow y' = y + b , \qquad y' \longrightarrow y = y' - b ,$$
$$z \longrightarrow z' = z + c , \qquad z' \longrightarrow z = z' - c .$$

The *rotation* transformation maps all points of space to image points found by rotation about a fixed common axis through a common angle. Rotations are invertible. If the axis of rotation is taken as the z axis and the rotation angle is α (in the positive sense, from the positive x axis to the positive y axis), we have [50, 51]

$$x \longrightarrow x' = x \cos \alpha - y \sin \alpha , \qquad x' \longrightarrow x = x' \cos \alpha + y' \sin \alpha ,$$
$$y \longrightarrow y' = x \sin \alpha + y \cos \alpha , \qquad y' \longrightarrow y = -x' \sin \alpha + y' \cos \alpha ,$$
$$z \longrightarrow z' = z , \qquad z' \longrightarrow z = z' .$$

The *plane reflection* (or *plane inversion* or *mirror reflection*) transformation is the transformation of reflection through a fixed two-sided plane mirror, the reflection plane. The image of any point is found by dropping a perpendicular from the point to the reflection plane and continuing the line on for the same distance on the opposite side of the plane. The image point is located at the end of the line segment.

Plane reflections are invertible. If the reflection plane is taken as the xy plane, we have

$$x \longrightarrow x' = x , \qquad x' \longrightarrow x = x' ,$$
$$y \longrightarrow y' = y , \qquad y' \longrightarrow y = y' ,$$
$$z \longrightarrow z' = -z , \qquad z' \longrightarrow z = -z' .$$

The transformation of *line inversion* (or *line reflection*) is inversion through a fixed straight line, the inversion line. The image of any point is the point at the end of the line segment constructed by dropping a perpendicular from the original point to the inversion line and continuing the perpendicular on for the same distance. Line inversions are invertible. If the inversion line is taken as the z axis, we have

$$x \longrightarrow x' = -x , \qquad x' \longrightarrow x = -x' ,$$
$$y \longrightarrow y' = -y , \qquad y' \longrightarrow y = -y' ,$$
$$z \longrightarrow z' = z , \qquad z' \longrightarrow z = z' .$$

By putting $\alpha = 180°$ in the rotation formulas above, we find that line inversion and rotation by $180°$ about the inversion line are the same transformation.

For the *point inversion* (or *point reflection* or *space inversion*) transformation the image of any point is at the end of the line segment running from the object point through a fixed point, the inversion center, and on for the same distance. Point inversions are invertible. If the inversion center is taken as the coordinate origin, we have

$$x \longrightarrow x' = -x , \qquad x' \longrightarrow x = -x' ,$$
$$y \longrightarrow y' = -y , \qquad y' \longrightarrow y = -y' ,$$
$$z \longrightarrow z' = -z , \qquad z' \longrightarrow z = -z' .$$

The *glide* transformation is the transformation consisting of the consecutive application of displacement parallel to a fixed plane and reflection through the plane, called a glide plane. (The two transformations can just as well be applied in reverse order, since they commute.) Glide transformations are invertible. If the glide plane is taken as the xy plane, we have

$$x \longrightarrow x' = x + a , \qquad x' \longrightarrow x = x' - a ,$$
$$y \longrightarrow y' = y + b , \qquad y' \longrightarrow y = y' - b ,$$
$$z \longrightarrow z' = -z , \qquad z' \longrightarrow z = -z' .$$

The *screw* transformation is the transformation resulting from the consecutive application (in either order, since they commute) of a rotation and a displacement parallel to the rotation axis, called a screw axis. Screw transformations are invertible. If the screw axis is taken as the z axis, we have for rotation through angle α along with displacement by c

$$x \longrightarrow x' = x \cos \alpha - y \sin \alpha , \qquad x' \longrightarrow x = x' \cos \alpha + y' \sin \alpha ,$$
$$y \longrightarrow y' = x \sin \alpha + y \cos \alpha , \qquad y' \longrightarrow y = -x' \sin \alpha + y' \cos \alpha ,$$
$$z \longrightarrow z' = z + c , \qquad z' \longrightarrow z = z' - c .$$

Under a *spatial dilation* (or *scale*) transformation the image of any point is found by moving the point away from a fixed point, the dilation center, along the straight line connecting them, to a distance from the fixed point related to the original distance by a fixed positive factor, the dilation factor. This transformation increases the distances between all pairs of points by the same factor. (If the dilation factor is less than 1, all distances are actually decreased.) Dilations are invertible. If the dilation center is taken as the coordinate origin, we have for positive ρ

$$x \longrightarrow x' = \rho x , \qquad x' \longrightarrow x = x'/\rho ,$$
$$y \longrightarrow y' = \rho y , \qquad y' \longrightarrow y = y'/\rho ,$$
$$z \longrightarrow z' = \rho z , \qquad z' \longrightarrow z = z'/\rho .$$

The image of any point under a *plane projection* transformation is a point in a fixed plane, the projection plane. It is the point of intersection of the perpendicular dropped from the original point to the projection plane. Projections are not invertible; they are neither one-to-one nor onto. If the projection plane is taken as the xy plane, we have

$$x \longrightarrow x' = x ,$$
$$y \longrightarrow y' = y ,$$
$$z \longrightarrow z' = 0 .$$

Under a *line projection* transformation the image of any point is a point on a fixed line, the projection line. The image is the point of intersection of the perpendicular dropped from the original point to the projection line. If the projection line is taken as the z axis, we have

$$x \longrightarrow x' = 0 ,$$
$$y \longrightarrow y' = 0 ,$$
$$z \longrightarrow z' = z .$$

We now look at three temporal transformations. All are invertible. The first is *temporal* (or *time*) *displacement* (or *translation*), under which the images of all instants are the same time interval d away from the object instants:

$$t \longrightarrow t' = t + d , \qquad t' \longrightarrow t = t' - d .$$

The transformation of *temporal* (or *time*) *inversion* (or *reflection* or *reversal*) maps each instant to the instant that is the same time interval before a fixed instant, the central instant, as the object instant is after it, or vice versa. If the inversion central instant is taken as the temporal origin, we have

$$t \longrightarrow t' = -t , \qquad t' \longrightarrow t = -t' .$$

The *temporal* (or *time*) *dilation* (or *scale*) transformation maps all instants to instants whose time intervals from a fixed instant, the central instant, are larger by a fixed positive factor, the dilation factor, than those of the respective object instants. The transformation increases the time intervals between all pairs of instants by the same factor. (If the dilation factor is less than 1, all time intervals are actually decreased.) If the dilation central instant is taken as the temporal origin, we have for positive σ

$$t \longrightarrow t' = \sigma t , \qquad t' \longrightarrow t = t'/\sigma .$$

Our first spatiotemporal transformation is the *Lorentz transformation* (Hendrik Antoon Lorentz, Dutch physicist, 1853–1928), also called *boost* or *velocity boost*, which maps all events to the events whose coordinates are the same as those that an observer moving with constant rectilinear velocity would assign to the original events with respect to his or her rest frame (the Minkowskian frame with respect to which he or she is at rest). Lorentz transformations are invertible. They form an essential ingredient of Albert Einstein's (German–American physicist, 1879–1955) special theory of relativity. If the observer is moving in the negative x direction with velocity v, such that $-c < v < c$, where c denotes the speed of light, and if his or her Minkowskian coordinate axes are parallel to ours and his or her origin coincides with ours at time $t = 0 = t'$, we have

$$x \longrightarrow x' = \gamma(x + vt) , \qquad x' \longrightarrow x = \gamma(x' - vt') ,$$
$$y \longrightarrow y' = y , \qquad y' \longrightarrow y = y' ,$$
$$z \longrightarrow z' = z , \qquad z' \longrightarrow z = z' ,$$
$$t \longrightarrow t' = \gamma(t + vx/c^2) , \qquad t' \longrightarrow t = \gamma(t' - vx'/c^2) ,$$

where

$$\gamma = \frac{1}{\sqrt{1 - v^2/c^2}} \, .$$

Our other spatiotemporal transformation is the *Galilei transformation* (Galileo Galilei, Italian astronomer, mathematician, and physicist, 1564–1642), also called *nonrelativistic boost* or *nonrelativistic velocity boost*. It is defined like the Lorentz transformation, except that the mathematical limit of $c \to \infty$ (or $v/c \to 0$) is taken. With the same assumptions as above we have

$$x \longrightarrow x' = x + vt \, , \qquad x' \longrightarrow x = x' - vt' \, ,$$
$$y \longrightarrow y' = y \, , \qquad y' \longrightarrow y = y' \, ,$$
$$z \longrightarrow z' = z \, , \qquad z' \longrightarrow z = z' \, ,$$
$$t \longrightarrow t' = t \, , \qquad t' \longrightarrow t = t' \, .$$

10.4 State Equivalence

In Sect. 4.2 we became acquainted with the notion of equivalence relation. The introduction there was a general one, not particularly for states of systems. Now, as part of the symmetry formalism that we are developing, we recall the definition of equivalence relation. This time it is expressed explicitly in terms of states and state space.

An equivalence relation for a state space of a system is any relation, denoted $\equiv$, that might hold between any pair of states that satisfies these three properties:

1. *Reflexivity.* Every state has the relation with itself, i.e.,

$$u \equiv u \, ,$$

   for all states u of the state space. Every state is equivalent with itself.

2. *Symmetry.* If one state has the relation with another, then the second has it with the first, for all states of the state space. In symbols that is

$$u \equiv v \quad \Longleftrightarrow \quad v \equiv u \, ,$$

   for all states u, v of the state space. If state u is equivalent with state v, then v is equivalent with u.

3. *Transitivity.* If one state possesses the relation with a second state
   and the second has it with a third, then the first state has the
   relation with the third, for all states of the state space. Represented
   symbolically, that is

$$u \equiv v, \quad v \equiv w \implies u \equiv w,$$

   for all states u, v, w of the state space. If state u is equivalent with
   state v and v is equivalent with state w, then u is equivalent with w.

Recall from Sect. 4.2 also that any subset of states – which in our
context is called a subspace of state space – such that all the states
it contains are equivalent with each other and with no other state, is
called an equivalence class, or, in our context, an equivalence subspace.
And recall also that an equivalence relation brings about a decomposition of a state space into equivalence subspaces.

Here are some examples of state equivalence:

1. If microstates of the gas of previous examples are being considered,
   while only the macroscopic properties of the gas are really of interest, any two microstates corresponding to the same pressure, volume, and temperature can be taken as equivalent. Actually, there
   is an infinite number of microstates corresponding to any given
   macrostate and thus forming an equivalence subspace. As for the
   infinity of microstates that do not correspond to any equilibrium
   macrostate, we can define all of them as equivalent to each other.

2. Considering only macrostates of the gas, for some purposes states
   with the same temperature, for example, might be equivalent. That
   would occur if the gas served solely as a heat sink. Then all states
   with the same temperature would form an equivalence subspace.

3. Take an equilateral triangle for the plane figure of given shape and
   size, lying in a given plane, with one of its points fixed in the plane.
   As far as appearance is concerned, any state of the triangle will be
   equivalent to the two states obtained from it by rotations by 120°
   and by 240° about the axis through its center (as the fixed point)
   and perpendicular to its plane and to the three states obtained
   from it by reflection through the three planes containing this axis
   and a median. Thus, the state space of the system decomposes into
   equivalence subspaces of six equivalent states each.

4. If the triangle of the preceding example has one vertex marked or
   if it is deformed into an isosceles triangle, then the states obtained
   from any given state by rotations by 120° and 240° will no longer

be equivalent to the original state with respect to appearance. And of the reflections, only the reflection through the plane containing the median of the different vertex and perpendicular to the plane of the triangle will still produce an equivalent state. Now the state space of the system decomposes into equivalence subspaces of only two states each.

5. If yet another vertex of the triangle is marked and differently from the first, or if the isosceles triangle is further deformed into a scalene triangle, then all equivalence among states, with regard to appearance, will disappear.

6. In the system of one ball and three depressions, if the three depressions are adjacent to one another and the scene is viewed from a sufficient distance, then as far as appearance is concerned all three states will look the same from that distance and will be equivalent.

7. Now imagine that only two of the depressions are adjacent to one another with the third separated from them, and view the system again from a distance. As far as appearance is concerned, two of the states will be equivalent with each other and the third will not be equivalent with either of them. The state space will then decompose into two equivalence subspaces, one containing two states and the other containing one.

8. If the same system is viewed from close up, no state will look like another in any case, and there will be no equivalence with regard to appearance.

9. For the system of three balls in three depressions, if all three balls look the same, than as far as appearance is concerned all six states of the system will be equivalent with each other.

10. In the preceding example if two balls look alike (are red, say) and the third is distinct (perhaps blue), each state of the system will be equivalent with the state obtained from it by interchanging the similar balls. Thus, the state space of six states will decompose into three equivalence subspaces of two states each.

11. If all three balls look different, there will be no equivalence among states with regard to their appearance.

12. Since quantum systems are especially important, we discuss quantum state equivalence and its consequences separately in Sect. 10.8.

10.5 Symmetry Transformation, Symmetry Group

One might observe that the symmetry formalism we are developing and our earlier, mostly conceptual discussion of symmetry are converging. A change is represented by a transformation, while transformation groups represent families of changes. An equivalence relation for a state space of a system brings about a decomposition of the state space into equivalence subspaces, which is classification, which implies analogy, which is symmetry (see Sects. 1.3 and 2.5).

The fundamental definition of symmetry as immunity to a possible change, which was presented in Sect. 1.1, now finds formal expression within the framework that is being erected in the present chapter. Any transformation (i.e., a replacement of states with states) that replaces states only with states that are equivalent to them leaves unchanged the property of states that is their membership in an equivalence subspace. In simpler language: As long as the equivalence relation is not the trivial one that every state is equivalent only with itself, it is always possible to make a nontrivial change, i.e., perform a transformation that is not the identity transformation, that replaces states with equivalent states only. Such a transformation operates solely within individual equivalence subspaces and does not mix them up. Thus, the property of states that they belong to a certain equivalence subspace is immune to such a transformation. That gives symmetry as the immunity of equivalence class membership to a possible transformation. Such a transformation is called a *symmetry transformation.*

And again, in the symmetry formalism we are developing, decomposition of a state space into nontrivial equivalence subspaces represents the aspect of the situation that is left unchanged by possible changes, which in our formalism are transformations. So symmetry is represented by the existence of transformations that leave equivalence subspaces invariant, i.e., transformations that map every state to an image state that is equivalent with the object state. Such a transformation is a symmetry transformation. Thus, the defining property of a symmetry transformation S is

$$u \xrightarrow{S} v \equiv u \quad \text{or} \quad S(u) = v \equiv u \,,$$

for all states u of the state space, where state v is the image of u.

The set of all invertible symmetry transformations of a state space of a system for an equivalence relation forms a group, a subgroup of the transformation group, called the *symmetry group* of the system for the equivalence relation. (Do not confuse symmetry group with symmetric group!) That is seen as follows:

1. Closure follows from transitivity of the equivalence relation. In Fig. 10.6 it is proved that if S_1 and S_2 are symmetry transformations, so is S_2S_1. The proof for S_1S_2 is similar.

2. Associativity holds for composition of transformations by consecutive application, as shown in Sect. 10.2.

3. The identity transformation is a symmetry transformation. That follows from reflexivity of the equivalence relation:

$$u \xrightarrow{I} u \equiv u \quad \text{or} \quad I(u) = u \equiv u \,,$$

for all states u.

4. The inverse of any invertible symmetry transformation is also a symmetry transformation. Let S be an invertible symmetry transformation, so that

$$u \xrightarrow{S} v \equiv u \quad \text{or} \quad S(u) = v \equiv u \,,$$

for all states u. Then by the symmetry property of the equivalence relation

$$v \xrightarrow{S^{-1}} u \equiv v \quad \text{or} \quad S^{-1}(v) = u \equiv v \,,$$

for all states v. Thus S^{-1} is indeed a symmetry transformation.

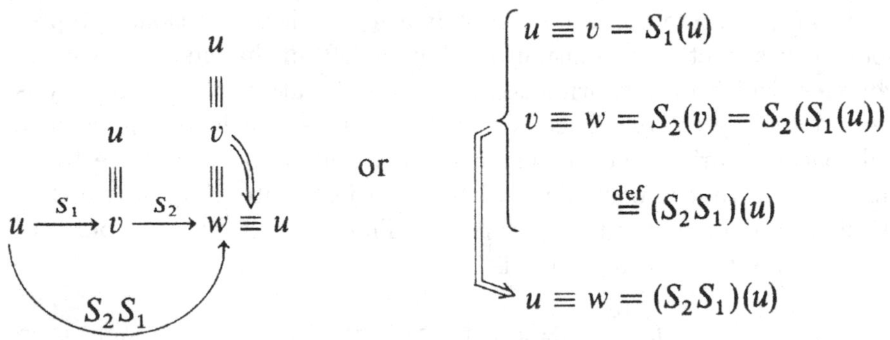

Fig. 10.6. Proof of closure for the composition of symmetry transformations by consecutive application: S_2S_1 is a symmetry transformation if S_1 and S_2 are. Arrow notation is shown on the left, and function notation on the right. The relations are valid for all states u of the state space. State v is the image of u by S_1, and state w is the image of v by S_2

So the set of all invertible symmetry transformations of a state space of a system for an equivalence relation possesses the four group properties and thus forms a group, a subgroup of the transformation group of the state space. In general a system might have different symmetry groups for its different state spaces and for the different equivalence relations that might be defined for them. If no two states are equivalent, i.e., if every state is equivalent only with itself, the symmetry group will consist only of the identity transformation, and the system will be asymmetric. In that case, although changes are possible, there are none that leave an aspect of the situation unchanged. In the other extreme, if all states are equivalent with each other so that all of state space forms a single equivalence subspace, the symmetry group will be the transformation group itself.

The line of reasoning by which any equivalence relation determines a subgroup of the transformation group can be reversed to allow any subgroup of the transformation group to determine an equivalence relation. Given such a subgroup, the corresponding equivalence relation is simply: State u is equivalent with state v if and only if some transformation T that is an element of the subgroup transforms u to v,

$$u \xrightarrow{T} v \quad \text{or} \quad v = T(u) .$$

That this relation is indeed an equivalence relation is seen as follows:

1. The reflexivity property of the relation follows from the identity transformation's belonging to the subgroup.
2. The symmetry property follows from the existence of inverses for the subgroup.
3. Transitivity follows from the subgroup's closure under composition by consecutive application.

Such a subgroup might or might not be the symmetry group for the equivalence relation it determines in that way. Clearly all elements of the subgroup are symmetry transformations for the equivalence relation it determines. There might, however, exist additional invertible symmetry transformations. If not, the subgroup is the symmetry group. In any case it is a subgroup of the symmetry group.

Consider some examples of symmetry groups:

1. In the case of a gas, where microstates corresponding to the same macrostate are considered equivalent to each other, the (infinite-order) symmetry group, transforming microstates to equivalent mi-

crostates only and thus preserving macrostates, is a subgroup of the (infinite-order) transformation group.

2. If macrostates of the gas having the same temperature are considered equivalent, we will have the constraint $T(p, V) = $ const. (isotherms), and all symmetry transformations will be of the form

$$p \longrightarrow p' = f(p, T = \text{const.}) \, ,$$

for arbitrary positive functions $f(p, T)$. Those transformations form a subset of all transformations of this state space as presented in Sect. 10.2. For arbitrary invertible positive functions $f(p, T)$ whose range of values is all positive numbers, we obtain the (infinite-order) symmetry group, a subgroup of the (infinite-order) transformation group.

3. The symmetry group of the equilateral triangle (see Sect. 10.4) is the group consisting of the identity transformation, rotations by 120° and by 240° about the axis through the center of the triangle and perpendicular to its plane, and reflections through each of the three planes containing the rotation axis and a median, the order-6 group D_3. Refer to Fig. 10.7 for the system and to Fig. 9.11 in Sect. 9.8 for the group table. That group is a subgroup of the transformation group of the system, which is the same as the

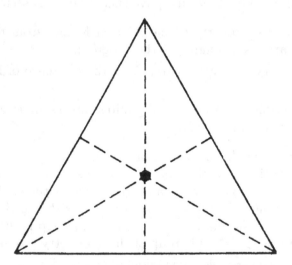

Fig. 10.7. Equilateral triangle. The center is the point of intersection of the axis of three-fold rotation symmetry, perpendicular to the page. The medians are lines of intersection of planes of reflection symmetry, also perpendicular to the page

transformation group of any plane figure of given shape and size, lying in a given plane, with one of its points fixed in the plane, and was presented in Sect. 10.2.

4. For the equilateral triangle with one vertex marked or for an isosceles triangle, the symmetry group consists only of the identity transformation and reflection through the single plane containing the median of the distinct vertex and perpendicular to the plane of the triangle, the order-2 group C_2. See Fig. 10.8 for the triangle and Fig. 8.2 in Sect. 8.1 for the group table. That group is a subgroup of the transformation group of the system and is also a proper subgroup of the symmetry group of the equilateral triangle.

5. For the equilateral triangle with two vertices marked differently or for a scalene triangle, the symmetry group consists only of the identity transformation. Those systems are asymmetric. Their trivial symmetry group is a subgroup of the transformation group of the system and is also a proper subgroup of the symmetry group of the singly marked equilateral triangle or the isosceles triangle and thus also a proper subgroup of the symmetry group of the equilateral triangle.

6. For a similar example, replace the equilateral triangle of the preceding examples with a square. The symmetry group of this system for the equivalence relation of identical appearance consists of the identity transformation, rotations by 90°, 180°, and 270° about the axis through the center of the square and perpendicular to its

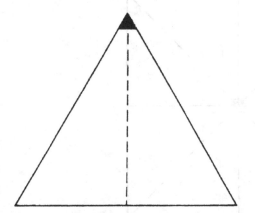
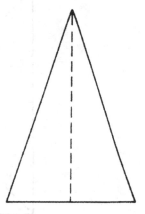

Fig. 10.8. Equilateral triangle with one vertex marked and isosceles triangle. The median of the distinct vertex is the line of intersection of the plane of reflection symmetry, perpendicular to the page

plane, reflections through each of the two planes containing the axis and a diagonal, and reflections through each of the two planes containing the axis and parallel to a pair of edges, the order-8 group D_4. Refer to Fig. 10.9 for the square and to Fig. 8.30 in Sect. 8.5 for the group table. That symmetry group is a subgroup of the transformation group of the system (the same as that of any plane figure of given shape, etc.).

Note that the symmetry group of the equilateral triangle is D_3 and that of the square is D_4. A pattern is emerging here. We take this opportunity to present the definition of the general *dihedral group* D_n. In a generalization of the symmetry groups of the equilateral triangle and of the square, the dihedral group D_n is the symmetry group of the regular n-sided polygon (also called regular n-gon). It consists of the identity transformation, rotations about the axis through the center of the polygon and perpendicular to its plane by $360°/n$, $2 \times 360°/n$, $3 \times 360°/n$, ..., $(n-1) \times 360°/n$, and reflections through each of the n planes containing the axis and a vertex or the center of a side (or both). D_n is of order $2n$.

7. If the square is squeezed into a rectangle, its symmetry group will reduce to the identity transformation, rotation by 180° about the axis through the center of the rectangle and perpendicular to its plane (or inversion through the axis, which is the same – see

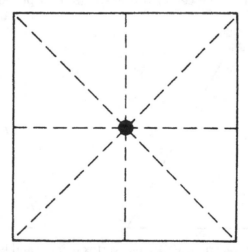

Fig. 10.9. Square. Lines of intersection of planes of reflection symmetry, perpendicular to the page, are indicated. The center is the point of intersection of the four-fold rotation symmetry axis, also perpendicular to the page

Sect. 10.3), and reflections through each of the two planes containing the axis and parallel to a pair of edges, the order-4 group D_2. Refer to Fig. 10.10 for the rectangle and to Fig. 8.5 in Sect. 8.1 for the group table. That group is a subgroup of the transformation group of the system (that of any plane figure of given shape, etc.) and is also a proper subgroup of the symmetry group of the square.

8. If for the system of one ball and three depressions all three depressions are adjacent so that all three states look the same from a sufficient distance, then all states will be equivalent, and the symmetry group will be the transformation group itself, the order-6 group S_3. That group happens to be isomorphic with the symmetry group of the equilateral triangle of given size, etc., D_3 (see Fig. 9.10 in Sect. 9.7).

9. If only two of the depressions are adjacent, so that only two of the three states are equivalent to each other, the symmetry group will consist of the identity transformation and the permutation interchanging the two equivalent states, the order-2 group S_2 (see Sect. 9.7). That group is a subgroup of the transformation group of the system, S_3, which is also the symmetry group of the system with all depressions adjacent. In addition, S_2 is isomorphic with

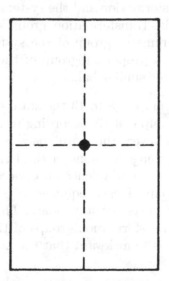

Fig. 10.10. Rectangle. Lines of intersection of planes of reflection symmetry, perpendicular to the page, are indicated. The center is the point of intersection of the two-fold rotation symmetry axis, also perpendicular to the page

the symmetry group of the isosceles triangle (or equilateral triangle with one vertex marked) of given size, etc., C_2 (see Sect. 8.1).

10. If all three depressions are separated and no states are equivalent, the symmetry group will consist only of the identity transformation. That is a subgroup of the transformation group and is also a proper subgroup of the symmetry group of the system with two adjacent depressions and thus also a proper subgroup of the symmetry group of the system with three adjacent depressions.

11. If in the system of three balls in three depressions all three balls look the same, then, since all states are equivalent, the symmetry group will be the transformation group itself, the order-720 group S_6 (see Sect. 9.7).

12. If only two balls look alike and the third is distinct, the symmetry group will be the direct product of the permutation groups of the individual equivalence subspaces, the direct product of the symmetric group of degree 2 with itself twice, $S_2 \times S_2 \times S_2$, an order-8 group. That group is a subgroup of the transformation group of the system, S_6, which is also the symmetry group of the system with three similar balls (see Sects. 9.6 and 9.7).

13. If all three balls look different, the symmetry group will consist only of the identity transformation and the system is asymmetric. That is a subgroup of the transformation group and is also a proper subgroup of the symmetry group of the system with two similar balls and thus also a proper subgroup of the symmetry group of the system with three similar balls.

Note that in each of examples 8 to 13 the state space is finite and we found the symmetry group exactly according to its definition. In each of examples 1 to 7, however, the state space is infinite (all orientations and both mirror image versions of the figure) and decomposes into an infinite number of equivalence subspaces. What we found is the transformation group of each equivalence subspace rather than the symmetry group of the entire state space. The latter is the infinite direct product of the transformation groups of the individual equivalence subspaces, which is so awkward that the normal procedure is to do just what we did.

Symmetry groups for quantum systems are discussed in Sect. 10.8.

10.6 Approximate Symmetry Transformation

In Sect. 1.1 we became acquainted with the concept of approximate symmetry, and in Sect. 6.1 we started formalizing the concept by introducing the notion of *approximate symmetry transformation* with the help of a *metric*. The idea of a metric was not developed beyond a conceptual description, which allowed a definition of approximate symmetry transformation that was sufficient for the purpose of Chap. 6. In the present section, in order to include approximate symmetry in the symmetry formalism we are developing, we will make the ideas more precise.

So, picking up from Sect. 6.1, we need to make precise the notion of *approximate symmetry transformation*, which is any transformation that changes every state of a system to a state that is 'nearly equivalent' to the original state. And just what does 'nearly equivalent' mean? For that we must soften the all-or-nothing character of the equivalence relation, upon which symmetry is based (see Sects. 10.4 and 10.5), in order to allow, in addition to equivalence, varying degrees of inequivalence. The way to do that is to define a *metric* for a set of states of a system, a 'distance' between every pair of states, such that null 'distance' indicates equivalence and positive 'distances' represent degrees of inequivalence.

More precisely, a metric is a nonnegative function of two states, denoted $d(\ ,\ )$, having the following properties for all states u, v, w of a state space of a system:

1. *Null self-distance* $d(u, u) = 0$,
2. *Symmetry* $d(u, v) = d(v, u)$,
3. *Triangle inequality* $d(u, w) \leq d(u, v) + d(v, w)$.

A metric is a generalization of an equivalence relation and includes it as a special case, where null distance between states u and v,

$$d(u, v) = 0 ,$$

defines equivalence of those states:

$$d(u, v) = 0 \quad \stackrel{\text{def}}{\Longrightarrow} \quad u \equiv v .$$

Null distance is indeed an equivalence relation, since the properties of (1) reflexivity, (2) symmetry, and (3) transitivity of an equivalence relation (see Sect. 10.4) follow, respectively, from the properties of (1) null self-distance, (2) symmetry, and (3) triangle inequality of a metric.

1. Null self-distance means, by definition, that every state is equivalent with itself, which is the reflexivity property of an equivalence relation.

2. By the symmetry property of a metric, if states u and v have null distance, then so do states v and u, for all states u and v. That, by definition, is the same as the statement that if u is equivalent with v, then v is equivalent with u, which is the symmetry property of an equivalence relation.

3. If states u and v have null distance and so do states v and w, then by the triangle inequality relation states u and w must have null distance, for all states u, v, w. That translates into the transitivity property of an equivalence relation, that if u and v are equivalent and v and w are equivalent, then u is equivalent with w.

For a set of states that is equipped with a metric $d(\ ,\ )$ some of the transformations might change all states to 'nearby' states only. To express that more precisely, we denote a general transformation by T and use function notation to indicate the result of changing state u by transformation T, which is then $T(u)$. For a given positive number ε there might be transformations T such that

$$d\big(u, T(u)\big) < \varepsilon\,,$$

for all states u. Such a transformation is called an *approximate symmetry transformation* of the system. We see that whether a given transformation is an approximate symmetry transformation can depend on the value of ε. The larger the value of ε, the more transformations are approximate symmetry transformations in general. Note that all the approximate symmetry transformations for a set of states of a system include among them all the exact symmetry transformations (see Sect. 10.5). That is because, for symmetry transformation S, which transforms all states to equivalent states, meaning all states u to states that have null distance with u, we have

$$d\big(u, S(u)\big) = 0 < \varepsilon\,.$$

So all symmetry transformations of a system are also approximate symmetry transformations for any value of ε.

The set of all invertible approximate symmetry transformations of a given system for any of its state spaces equipped with a metric and for a given value of ε does not in general form a group. There might be trouble with closure. The set does include the corresponding symmetry group, as we saw in the preceding paragraph, and might include other groups that in turn include the symmetry group.

For our purpose in this book it is sufficient to mention here that it is nevertheless possible to define *approximate symmetry groups* for state spaces equipped with metrics, and it is possible to define a measure of *goodness of approximation* for every approximate symmetry group. See [52], for example. A system will in general possess different approximate symmetry groups with different measures of goodness of approximation. For the same state space and metric, every approximate symmetry group will include the corresponding symmetry group as a subgroup. It might also include as subgroups approximate symmetry groups of higher goodness of approximation than that of itself.

An approximate symmetry group does not in general contain all the invertible approximate symmetry transformations for some value of ε. And perhaps somewhat strangely, for any given value of ε an approximate symmetry group might contain transformations that are not approximate symmetry transformations at all.

In Sect. 6.1 we became acquainted with the ideas of *exact symmetry limit, broken symmetry*, and *symmetry breaking factor*. We will not elaborate on them further here.

10.7 Quantification of Symmetry

We now have at our disposal the means to attempt to quantify symmetry or at least set up a hierarchy, or ordering, of symmetries. A quantification of symmetry would be a way of assigning a number to each symmetry group, expressing the degree of symmetry of a system possessing that symmetry group. A symmetry ordering would be a way of comparing any two symmetry groups to determine which of two systems, each possessing respectively one of the two symmetry groups, has the higher degree of symmetry. A symmetry quantification is clearly a symmetry ordering, although an ordering might not go so far as to be a quantification.

There are three properties that, on the basis of our experience, we may reasonably expect of any scheme of symmetry quantification or ordering. First of all, if the symmetry group of a system is isomorphic with the symmetry group of another system (e.g., the equilateral triangle of given size, etc., and the ball and three adjacent depressions, or the isosceles triangle and the ball and two adjacent and one separated depressions), it is reasonable to consider them as possessing the same degree of symmetry (even though their symmetry transformations might be of entirely different character, as in the examples). Thus, it is the abstract group of which the symmetry group is a realization that is of interest, rather than the symmetry group itself.

Next, if the symmetry group of a system is isomorphic with a proper subgroup of the symmetry group of another system (e.g., the isosceles triangle and the equilateral triangle, or the ball and two adjacent and one separated depressions and the ball and three adjacent depressions, or the isosceles triangle and the ball and three adjacent depressions, or the rectangle and the square), the latter system may reasonably be considered more symmetric than the former.

And finally, if a system is asymmetric – if its symmetry group consists solely of the identity transformation – it is reasonable to assign it the lowest degree of symmetry.

But how should we compare the degrees of symmetry of systems whose symmetry groups are neither isomorphic with each other nor one isomorphic with a proper subgroup of the other (e.g., the square and the equilateral triangle)? The main purpose of any symmetry quantification or ordering scheme is to answer just that question.

One possibility for a quantification scheme is to take for the degree of symmetry of a system the order of its symmetry group (or any monotonically ascending function thereof, such as its logarithm). For finite-order symmetry groups (and we do not go into infinite-order groups here) that is a perfectly satisfactory scheme in that it possesses the three properties that we demanded above: Isomorphic groups have the same order, a proper subgroup is of lower order than its including group, and the order-1 group has the lowest order of all. However, it might well be objected on philosophical grounds that this quantification scheme assigns equal weights to all the elements of a symmetry group, i.e., to all the symmetry transformations of a system, while in fact they are not all independent, as they can be generated by repeated and consecutive applications of a minimal set of generators.

We are thus led to consider the scheme whereby the degree of symmetry of a system is taken to be the minimal number of generators of its symmetry group. As for the three properties, it possesses the first: Isomorphic groups have the same minimal numbers of generators. It is, however, deficient with respect to the second property: a proper subgroup does not necessarily have a smaller minimal number of generators than does its including group. For example, the order-2 cyclic group C_2 is a proper subgroup of the order-4 cyclic group C_4, yet, be-

Table 10.1. Table of orders and minimal numbers of generators for various symmetry groups

System	Symmetry group	Order	Minimal number of generators
Square	D_4	8	2
Rectangle	D_2	4	2
Equilateral triangle	$D_3 \sim S_3$	6	2
Isosceles triangle	$C_2 \sim S_2$	2	1
Ball and three adjacent depressions	$S_3 \sim D_3$	6	2
Ball and two adjacent and one separated depressions	$S_2 \sim C_2$	2	1
Asymmetric system	C_1	1	0 (defined)

ing cyclic, each group can be generated by positive powers of a single element. The third property can be forced by assigning the value 0 to the order-1 group.

For comparison we list in Table 10.1 some of the symmetries we found in our examples along with the degrees of symmetry assigned to the systems by the two symmetry quantification schemes that we have considered. Note, for example, that the minimal-number-of-generators scheme assigns equal degrees of symmetry to the square and to the rectangle. So the group order scheme is clearly preferable in that regard. Other, intermediate schemes might be devised, giving the degree of symmetry of a system as some function of both the order and the minimal number of generators of its symmetry group.

10.8 Quantum Systems

This section is a concise presentation, intended to place the subject of symmetry in quantum systems within the formalism developed in this chapter. Thus, it should introduce, but by no means replace, the various discussions of the same subject found elsewhere. It is intended for readers with a good understanding of the Hilbert space formulation of quantum theory. Others are advised to skip this section.

Quantum systems, which are but special cases of the general systems we have been discussing, are so important that they deserve separate treatment. Their special nature allows us to be more specific in our discussion of their symmetry.

The state spaces of quantum systems are Hilbert spaces, and quantum states are vectors in those spaces. (Actually, quantum states are *represented* by vectors in Hilbert spaces. More accurately, they are represented by *rays*, since all complex multiples of any vector are supposed to represent the same state. But for the convenience of the present discussion we identify state with vector. For the sake of simplicity we also ignore possible complications due to coherent subspaces, which in any case can be overcome with slight additional effort.) The transformations of interest for quantum systems are linear and antilinear transformations in their Hilbert spaces, said to be implemented by linear and antilinear operators.

Recall that a linear operator L in a Hilbert space is characterized by the properties

$$L(|u\rangle + |v\rangle) = L|u\rangle + L|v\rangle ,$$

$$Lz|u\rangle = zL|u\rangle ,$$

for all vectors $|u\rangle$, $|v\rangle$ and all complex numbers z. Similarly, an antilinear operator A obeys

$$A(|u\rangle + |v\rangle) = A|u\rangle + A|v\rangle ,$$

$$Az|u\rangle = \bar{z}A|u\rangle ,$$

where $\bar{z}$ denotes the complex conjugate of z.

Now, what properties of states of quantum systems might be utilized for the definition of equivalence relations? Consider the following:

1. *Transition amplitudes.* Every state $|u\rangle$ is characterized by the transition amplitudes between itself and all states of its Hilbert space, $\langle u|v\rangle$ or $\langle v|u\rangle$ for all states $|v\rangle$.

2. *Norm.* Especially, every state $|u\rangle$ is characterized by its transition amplitude with itself, $\langle u|u\rangle$, the square of its norm.

3. *Eigenheit.* Every state $|u\rangle$ either is or is not an eigenstate of each member of any set of operators $\{O_i\}$ in its Hilbert space, and if it is, it is characterized by its eigenvalue.

4. *Expectation values.* Every state $|u\rangle$ gives a set of expectation values $\{\langle u|O_i|u\rangle / \langle u|u\rangle\}$ for any set of operators $\{O_i\}$ in its Hilbert space.

It seems reasonable to define state equivalence for a given quantum system and a given set of operators in its Hilbert state space as indistinguishability with respect to all four properties.

The search for all symmetry operators for the equivalence relation of indistinguishability with respect to (1) transition amplitudes and (2) norm is essentially reducible to the search for all operators preserving scalar products up to complex conjugation. And the standard result is the group of all unitary and antiunitary operators in the Hilbert state space of the system [53].

As a reminder, unitary operators are linear operators U that obey

$$\langle u|U^\dagger U|v\rangle = \langle u|v\rangle \,,$$

for all states $|u\rangle$, $|v\rangle$, where the dagger denotes Hermitian conjugation. Antiunitary (or unitary antilinear) operators are invertible antilinear operators A obeying

$$\langle u|A^\dagger A|v\rangle = \langle v|u\rangle = \overline{\langle u|v\rangle} \,,$$

for all states $|u\rangle$, $|v\rangle$, where the bar denotes complex conjugation.

Now, consider property 3. Equivalent states, in addition to their being related by unitary or antiunitary operators, are also required to be either all eigenstates or all not eigenstates of each operator O_i of a given set $\{O_i\}$, and if they are all eigenstates, they must have the same eigenvalue. The first part of the requirement leaves as symmetry operators only those unitary and antiunitary operators commuting with the set $\{O_i\}$. The second part in general rejects the antiunitary operators among those, since, while eigenstates related by unitary operators indeed have the same eigenvalue of an operator commuting with the unitary operators, a pair of eigenstates related by an antiunitary operator commuting with the operator of which they are eigenstates have eigenvalues that form a complex-conjugate pair. However, if the set of operators $\{O_i\}$ consists solely of Hermitian operators, which indeed is the usual case, with Hermitian operators representing measurable physical quantities, the antiunitary operators are *not* rejected. In that case a pair of eigenstates related by an antiunitary operator commuting with the Hermitian operator of which they are eigenstates have the same eigenvalue, since the eigenvalues of Hermitian operators are real. A very important Hermitian operator that is almost always included in the set $\{O_i\}$ is the Hamiltonian operator (William Rowan Hamilton, British mathematician, 1805–1865), which represents the energy of the system and generates the system's evolution.

That also takes care of property 4. States that are related by unitary operators commuting with a given set of operators $\{O_i\}$ give equal sets of expectation values. A pair of states related by an antiunitary operator commuting with the set give complex-conjugate sets of expectation values. But if $\{O_i\}$ is a set of Hermitian operators, they will give equal sets of (real) expectation values.

In summary, then, for a given quantum system and a given set of operators in its Hilbert space of states, the symmetry group for equivalence with respect to (1) transition amplitudes, (2) norm, (3) eigenheit, and (4) expectation values is the group of all unitary operators in the Hilbert space commuting with the given set of operators. If the given set consists solely of Hermitian operators, the symmetry group is the group of all unitary and antiunitary operators commuting with the set. (But if it is desired to preserve scalar products strictly, not just up to complex conjugation, then the antiunitary operators are excluded whether the operators of the given set are Hermitian or not.)

10.9 Summary

In this chapter we developed a general symmetry formalism needed for the application of symmetry considerations in science, especially quantitative applications. In Sect. 10.1 we introduced these very general concepts: system, which is whatever we investigate the properties of; subsystem, a system wholly subsumed within a system; state of a system, a possible condition of the system; and state space of a system, which is the set of all states of the same kind. The concept of transformation, a mapping of a state space of a system into itself, was presented in Sect. 10.2. We saw that the set of all invertible transformations of a state space of a system forms a group, called a transformation group of the system.

Section 10.3 was a compilation of a number of transformations in space, in time, and in space-time. The spatial transformations presented were: displacement, rotation, plane reflection, line inversion, point inversion, glide, screw, dilation, plane projection, and line projection. The temporal transformations were displacement, inversion, and dilation. And the spatiotemporal transformations we saw were the Lorentz and Galilei transformations.

In Sect. 10.4 we considered the possibility of an equivalence relation for a state space of a system. Such a relation decomposes a state space into equivalence subspaces. That led in Sect. 10.5 to the idea of a symmetry transformation, which is any transformation that maps every state to an image state that is equivalent to the object state, i.e., any transformation that preserves equivalence subspaces. The set of all invertible symmetry transformations of a state space for an equivalence relation forms the symmetry group of the state space for the equivalence relation, and is a subgroup of the transformation group.

We brought approximate symmetry into the symmetry formalism in Sect. 10.6, where we made precise the notion of approximate symmetry transformation by means of a metric in the state space of a system, and we saw properties of metrics. Quantification of symmetry was discussed in Sect. 10.7, where we found that the order of a (finite-order) symmetry group, or any monotonically increasing function thereof, can reasonably serve as the degree of symmetry of a system possessing that symmetry group.

Our discussion of state equivalence for quantum systems in Sect. 10.8 led to the result that for a given quantum system and a given set of operators in its Hilbert space of states, the symmetry group is the group of all unitary operators commuting with the given set. If the given set consists solely of Hermitian operators, the symmetry group is the group of all unitary and antiunitary operators commuting with the set.

11

Symmetry in Processes

In Sect. 2.5 we discussed the reduction of the natural evolution of quasi-isolated systems into initial state and evolution. Then, in Sects. 3.1 and 3.2, we considered symmetry of evolution and of states. In the present chapter we will elaborate on those ideas and incorporate them into the framework of the symmetry formalism that we developed in Chap. 10.

We will start with a discussion of symmetry of evolution, also known as symmetry of the laws of nature. That will include an analysis of time reversal symmetry. Then we will consider symmetry of initial and final states of processes, leading to the equivalence and symmetry principles for natural processes in quasi-isolated systems and to the general and special symmetry evolution principles for such systems. The latter two principles are both concerned with the nondecrease of degree of symmetry during the evolution of quasi-isolated systems. We will obtain an explanation for the empirical observation that macrostates of stable equilibrium of a physical system are often especially symmetric. Symmetry and entropy will be shown to be related to each other.

11.1 Symmetry of the Laws of Nature

Laws of nature, or *laws of evolution*, are what in Sect. 4.1 we called causal relations in systems. But in the present context the term 'system', which in Chaps. 4 and 10 was intentionally left very vague and general, refers specifically to natural processes of quasi-isolated systems. The whole evolution process is the 'system'. Initial and final states of the process are 'parts' of that system, are subsystems of it. In fact, for lawful behavior of quasi-isolated systems the initial state is the 'cause subsystem' and the final state the 'effect subsystem', as was explained in Sects. 2.5, 3.1, and 4.1.

Please note well and beware of the two senses in which the term 'system' is being used here! One sense is that of a physical system, which can be described in terms of initial and final states and natural evolution from the former to the latter. The other sense is 'whatever we investigate the properties of', according to the presentation of Sect. 10.1. And here we are investigating the properties of the evolution processes of quasi-isolated systems (in the former sense), where those processes involve initial states evolving into final states. So in the latter sense of the term 'system', the initial state, the final state, and the evolution are all parts of the system that is the process.

Yes, it is the dynamic process, the temporal evolution of the physical system, that is taken as the 'system', while the state of the physical system at any initial time is the 'cause subsystem' and the state at any final time is the 'effect subsystem'. The laws of nature, or laws of evolution, are the causal relation between cause and effect subsystems in such cases, the causal relation between initial and final states of physical processes. Equivalently, and in more familiar language, the laws of nature can be viewed as the natural temporal development of physical systems from initial states to final states [54, 55].

It must be emphasized that we are considering only quasi-isolated physical systems, where by 'quasi-isolated' we mean that there exists minimal interaction with the rest of the world, that the physical systems evolve, to the extent possible, under internal influences only. That is implied by the existence of causal relation, cause subsystem, and effect subsystem, with initial states leading to unique final states.

Earlier in this book and possibly from other sources as well, you have read and heard of various symmetries and approximate symmetries that are ascribed to the laws of nature. They include such as special relativistic symmetry, SU(3) symmetry, charge symmetry, particle-antiparticle conjugation (C) symmetry, space inversion (P) symmetry, time reversal (T) symmetry, CP symmetry, and CPT symmetry. There are two points of view about the meaning of symmetry of the laws of nature:

1. A scientist (with a laboratory) investigates nature and discovers laws. Another scientist (with her laboratory), related to the first by some transformation, also investigates nature and discovers laws. If these laws are the same as those discovered by the first scientist (or if a certain subset of these is the same as a certain subset of those),

and if, moreover, that is true for all pairs of scientists related by the transformation, then the transformation is a symmetry transformation of the laws of nature (or of a certain subset of them). This is called the passive point of view.

2. If among all processes that can be conceived as occurring naturally (or a certain subset of them) any pair related by a certain transformation are either both allowed or both forbidden by nature (or, in quantum phenomena, both have the same probability), then that transformation is a symmetry transformation of the laws of nature (or of a certain subset of them). This is called the active point of view.

For example, the symmetry of the special theory of relativity can be expressed passively: (1) All pairs of scientists with arbitrary spatiotemporal separation, arbitrary relative orientation, and arbitrary constant rectilinear relative velocity discover the same laws of nature, including the speed of light. Or it can be put actively: (2) Among all processes that can be conceived as occurring naturally, including the propagation of a light signal over a certain distance during a certain time interval, any two that are identical except for arbitrary spatiotemporal separation, arbitrary relative orientation, and arbitrary constant rectilinear relative velocity are either both allowed or both forbidden by nature (or have the same probability).

Another example: CP symmetry, which seems to be valid in all of nature except for a certain class of weak interactions among elementary particles, can be formulated passively: (1) All pairs of scientists who are particle-antiparticle conjugates and space-inversion images of each other discover the same laws of nature (with certain exceptions). The transformation involved here is converting a scientist into an antiscientist, i.e., replacing all the protons, neutrons, and electrons composing the scientist and his laboratory with antiprotons, antineutrons, and positrons, respectively, and then point inverting the antiscientist (or first invert and then conjugate; the results are the same). Since antiscientists are not at present physically realizable (at least not in the sense just defined), the passive point of view seems absurd. So one might prefer the active one: (2) Among all processes that can be conceived as occurring naturally (with certain exceptions), any two that are identical except for particle-antiparticle conjugation and space inversion are either both allowed or both forbidden by nature (or have the same probability).

The two points of view are called passive and active, respectively, because of the nature of the transformations under which the laws of nature are symmetric. A transformation is called passive if it does not change the world but changes the way the world is observed, i.e., it changes the reference frame. For a spatial transformation, for example, that would be a change of coordinate system. A transformation is called active if it does not change the observer but changes the rest of the world. The same reference frame is used for a changed world. In the spatial case the locations of objects and events would be changed with respect to a fixed coordinate system. Refer also to our discussion of passive and active transformations in Sect. 3.3.

I prefer the active (second) point of view for considering symmetry of the laws of nature. The main reason is that, as we saw in the example concerning CP symmetry, transformed reference frames (observers) are not always physically realizable. From the active point of view, a symmetry transformation of the laws of nature can be concisely defined, just as in Sect. 3.1, as a transformation for which the transformed result of an experiment is the same as the result of the transformed experiment for all experiments (or for some subset of experiments, in which case we have a symmetry transformation of a subset of the laws of nature).

I will now present a precise, diagrammatic formulation of symmetry of the laws of nature. It is the expression, in the symmetry formalism we are developing, of the qualitative formulation presented in Sect. 3.1. The laws of nature, or laws of evolution, are expressed by a transformation, denoted N (for 'nature'), of the state space of every quasi-isolated physical system into itself, a mapping giving as the image of any state the final state evolving naturally from the object state as initial state and thus exhibiting the causal relation between initial and final states (see Fig. 11.1). Since not all states are necessarily obtainable as final states, there may exist states that do not serve as images in this mapping. How then, you might well ask, are such states obtained to serve as initial states of processes of natural evolution? They are set up by tampering with the system. They are the final states of processes during which the system is *not quasi-isolated*.

So, if u denotes the initial state of any physical system, $N(u)$ is the final state that is the result of the evolution of the system from initial state u. We can also use the notation

$$u \xrightarrow{N} N(u)$$

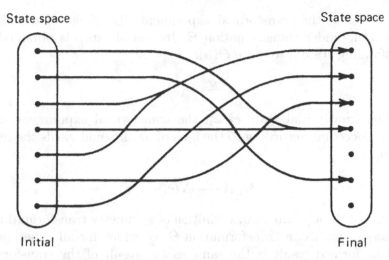

State space   State space

Initial   Final

Fig. 11.1. Laws of evolution for a quasi-isolated physical system are a mapping of the system's state space into itself, where the image of any state is the final state evolving from the object state as initial state

to make the transformation graphically look more like a process. The discrete, initial $\longrightarrow$ final model of evolution will best serve the purposes of our discussion. Continuous evolution is obtained in the limit of infinite reiteration of infinitesimal discrete steps.

Let Θ denote any transformation of every physical system, such as rotation, spatial displacement, reflection, or particle-antiparticle conjugation. The image of state u of any physical system is state $\Theta(u)$ of the same system, or

$$u \xrightarrow{\;\Theta\;} \Theta(u)\,,$$

although we are not indicating a process this time.

The process

$$u \xrightarrow{\;N\;} N(u)$$

is (or could be) the running of an experiment, where u denotes the initial experimental setup and $N(u)$ denotes the result. Transforming the experimental result $N(u)$ by Θ gives the image state $\Theta N(u)$, or

$$N(u) \xrightarrow{\;\Theta\;} \Theta N(u)\,.$$

Now, consider the transformed experiment, the image of the above experiment under transformation Θ. Its initial setup is obtained by transforming state u to state $\Theta(u)$, or

$$u \xrightarrow{\Theta} \Theta(u) .$$

Starting from initial state $\Theta(u)$, the transformed experiment takes place, proceeding according to the laws of nature, and yields the result $N\Theta(u)$, or

$$\Theta(u) \xrightarrow{N} N\Theta(u) .$$

Combining those results, our definition of symmetry transformation of the laws of nature, a transformation Θ for which for all experiments the transformed result is the same as the result of the transformed experiment (see Sect. 3.1) now becomes

$$\Theta N(u) = N\Theta(u) ,$$

for all states u, or diagrammatically as in Fig. 11.2. Since that holds for all states u, we have the formal, mathematical definition of a symmetry transformation of the laws of nature: The transformation under consideration commutes with the evolution transformation (see Sect. 10.2),

$$\Theta N = N\Theta .$$

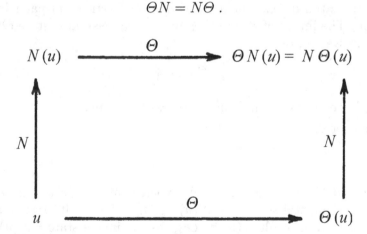

Fig. 11.2. A symmetry transformation of the laws of nature is any transformation Θ for which the diagram is valid for all states u. N is the evolution transformation

(For a symmetry transformation of a subset of the laws of nature, things must be formulated accordingly in terms of all states u belonging to a certain subspace of state space.)

The fundamental point underlying those definitions and formulations, as discussed in Sect. 3.1, is this: Symmetry of the laws of nature is indifference of the laws of nature. For a transformation to be a symmetry transformation of the laws of nature, the latter must ignore some aspect of physical states, and the transformation must affect that aspect only. A pair of initial states related by such a transformation are treated impartially by the laws of nature, so that they evolve into a pair of final states that are related by precisely the same transformation. The laws of nature are blind to the difference between the two states, which is then preserved during evolution and re-emerges as the difference between the two final states.

As an example, consider spatial-displacement symmetry of the laws of nature, meaning that the laws of nature are the same everywhere. If we perform two experiments that are the same except for one being here and the other being there, they will yield outcomes that are the same except for one being here and the other there, respectively. And that is found to be valid for all experiments and for all heres and theres. For a picture see Fig. 11.3.

Those insensitivities of the natural evolution of quasi-isolated systems remind us of the 'impotences' of scientific laws, discussed in Sect. 4.3. Indeed, since laws are expressions of nature's order, all sym-

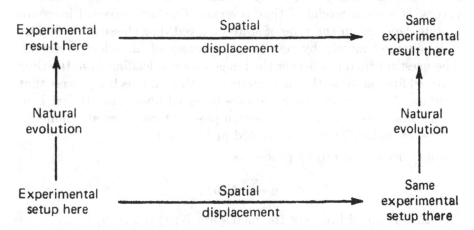

Fig. 11.3. Spatial-displacement symmetry of the laws of nature. The diagram is valid for all experiments and for all heres and theres

metries of evolution must appear as impotences of laws, if the laws are to express nature's order faithfully. Any subset of natural evolution processes will possess the symmetry of the full set and might have additional symmetry. Partial laws, those laws that are concerned with only part of the phenomena of nature, might accordingly possess additional impotences.

For example, since natural evolution is symmetric under spatial displacements, no scientific law may accept absolute position as input. For examples of partial theories, the gas laws are concerned with macroscopic states of gases, and Kirchhoff's rules are concerned with electric currents. In the former case, although the laws of nature do have something to say about which gas molecules go where and when, the gas laws ignore microscopic aspects of the gas. In the latter case, although the laws of nature do apply to the motion of each of the moving electrons that form the electric current, Kirchhoff's rules ignore such details.

The set of invertible symmetry transformations of the laws of nature, i.e., the set of all invertible transformations commuting with the evolution transformation, is easily shown to form a group. We call that group *the symmetry group of the laws of nature*.

Another possible symmetry of the laws of nature, very different in kind from the others, is temporal-inversion, or time reversal, symmetry, also called reversibility. The time reversal transformation acts on a process by replacing each state with its time reversal image and reversing the temporal order of events. A moving picture shown in reverse is a good model of time reversal. The time reversal image of a state depends on the type of state involved. For classical mechanics it is obtained merely by reversing the senses of all velocity vectors. The question then is whether the image process, leading from the time reversed final state to the time reversed initial state, is the process that would evolve naturally from the time reversed final state. If that is so for all processes, the laws of evolution possess time reversal symmetry, or are reversible. That is illustrated in Fig. 11.4.

In symbols, the object process is

$$u \xrightarrow{\ N\ } N(u) \ .$$

The time reversal image of the final state $N(u)$ is $TN(u)$,

$$N(u) \xrightarrow{\ T\ } TN(u) \ ,$$

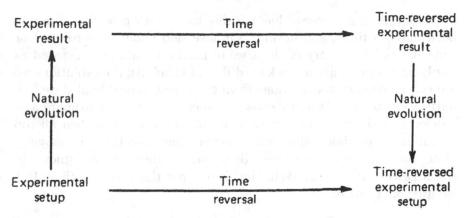

Fig. 11.4. Time reversal symmetry, or reversibility, of the laws of nature, if the diagram is valid for all experiments

where T denotes the time reversal transformation. The natural evolution from state $TN(u)$ is

$$TN(u) \xrightarrow{N} NTN(u) \ .$$

And if the final state of this process is the time reversal image of the initial state of the object process, i.e., if

$$NTN(u) = T(u) \ ,$$

for all states u, we have time reversal symmetry, the definition of which is then

$$NTN = T \ .$$

Note how this differs from the definition, derived previously, of a symmetry transformation of the usual kind Θ:

$$\Theta N = N\Theta \ .$$

At present, time reversal symmetry, or reversibility, seems to be valid in nature for the laws of evolution of almost all microscopic systems. (Microscopic systems containing neutral kaons form a notable exception.) On the other hand, that symmetry is invalid for almost all macroscopic systems. As a typical example, whereas gas flows spontaneously from a higher-pressure region to one of lower pressure, gas is never found to flow in the opposite direction spontaneously. Nor does a broken egg ever collect itself together and become whole again.

Note that time reversal has nothing to do with processes 'running backwards in time', and nothing we have said about time reversal or time reversal symmetry of the laws of nature should be construed as implying anything like a 'backward flow of time'. All the evolutions we are considering, even evolutions from the time reversed final state, are natural evolutions. True, we often represent time by a coordinate axis labeled t, and that is indeed very useful. However, it is then all too tempting to spatialize time and imagine that everything is 'moving' along that axis in the positive t direction. So then why not 'move' in the negative t direction? Well, that is just not the nature of time. But we are going astray.

11.2 Symmetry of Initial and Final States, the General Symmetry Evolution Principle

In the analysis of the behavior of quasi-isolated systems into initial state and evolution, the laws of evolution, or laws of nature, do not exhaust all of what is going on. The laws of nature do indeed determine how any process, once started, will evolve and what its outcome will be. But it is the initial state, the situation at any given single instant, that determines just which process is to take place. And, with an important exception discussed in Sect. 7.1, the laws of nature do not determine initial states. Or perhaps we might say that initial states are whatever nature allows us to have control over, at least in principle, or perhaps even better, whatever nature prefers not to be bothered with.

For an example in classical mechanics, consider a set of bodies interacting only with each other. The laws of evolution are Newton's laws of motion and the forces among the bodies. We may arbitrarily specify the positions, velocities, orientations, and angular velocities of all the bodies at a given time, but no more than that. So those specify initial states.

Since the laws of nature are beyond our control, the initial state uniquely determines for us the process and its outcome – the final state of the quasi-isolated system. Thus, as described in Sect. 11.1, initial state and outcome are in causal relation. Taking the whole process as our 'system', the initial state and the final state are cause and effect subsystems, respectively.

I again emphasize that we are considering only quasi-isolated physical systems, in which initial states do indeed uniquely determine the outcomes of processes, in which there is a minimum of outside influences messing things up.

Refer back to Sects. 3.2 and 4.3. Symmetry of the laws of nature (time reversal symmetry excluded) determines an equivalence relation in the state space of a system: A pair of states is equivalent if and only if they are indistinguishable by the laws of nature, i.e., if and only if some symmetry transformation of the laws of nature carries one into the other. Thus, the symmetry group of the laws of nature determines a decomposition of state space into equivalence subspaces, the physical significance of which is that the members of each equivalence subspace are all those states, and only those, that are indistinguishable by the laws of nature. Then, following reasoning similar to that in Sect. 4.3, we obtain the *equivalence principle for processes* in quasi-isolated systems (see Fig. 11.5):

> *Equivalent states, as initial states,* must *evolve into equivalent states, as final states, while inequivalent states* may *evolve into equivalent states.*

That principle is valid and useful for natural evolution processes, for any subset of them, for universal scientific laws, or for partial laws.

For an example of the equivalence principle for processes, since the laws of nature possess spatial-displacement symmetry, initial states that differ only in position evolve into final states that differ only in position.

Or, all known laws of nature are symmetric under interchange of elementary particles of the same species, so such particles are, as far as we know, inherently indistinguishable. Two initial states differing only in the interchange of, say, two electrons cannot, by any known process, lead to distinguishable final states.

Or, certain laws of nature might apply to macroscopic states of a system and be indifferent to microstates, as long as the latter correspond to the same macrostate. Then two different initial microstates corresponding to the same initial macrostate will evolve into final microstates corresponding to the same macrostate. Among those are the gas laws. Kirchhoff's rules care nothing about the composition and structure of resistors and emf (voltage) sources, as long as they have the resistance and emf they are supposed to have. Newton's laws are obviously indifferent to many aspects of the states to which they apply, such as odor, color, etc.

State space   State space

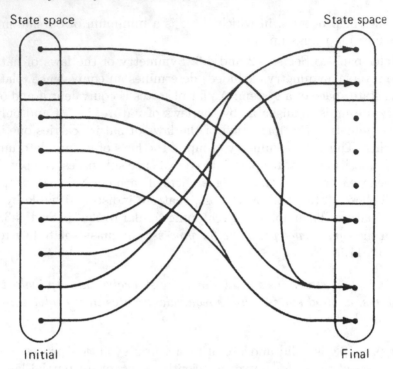

Initial   Final

Fig. 11.5. Decomposition of the state space of a quasi-isolated physical system into equivalence subspaces determined by the symmetry group of the laws of nature. According to the equivalence principle, equivalent states must evolve into equivalent states, while inequivalent states may also do so

Let us call the state space equivalence relation defined by the symmetry group of the laws of nature 'initial equivalence' for convenience. That really is for convenience in the following discussion, and is not just to use two words where one will do. We now define another, possibly different, equivalence relation, which we call 'final equivalence', as follows. A pair of states is final-equivalent if and only if the pair of states evolving from them is initial-equivalent, i.e., is equivalent with respect to the laws of nature. Initial equivalence implies final equivalence, by the equivalence principle for processes, since initial-equivalent states must evolve into initial-equivalent states and are thus final-equivalent. However, states that are not initial-equivalent may also evolve into initial-equivalent states and thus be final-equivalent. So, according to the equivalence principle for processes, in the decomposition of the state space of a physical system into final-equivalence subspaces, each final-equivalence subspace consists of one or more initial-equivalence subspaces in their entirety (see Fig. 11.6).

State space   State space

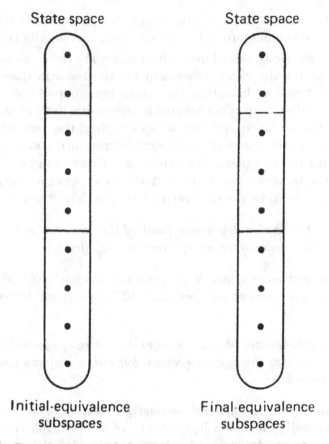

Initial-equivalence   Final-equivalence
subspaces   subspaces

Fig. 11.6. Decomposition of the state space of a quasi-isolated physical
system into initial-equivalence subspaces and into final-equivalence subspaces
for the example of Fig. 11.5. Each final-equivalence subspace consists of one
or more initial-equivalence subspaces

When considering symmetry in processes, the process is the 'system' in the general sense of the symmetry and equivalence principles, and the set of all possible processes of a physical system is the 'state space' in the same general sense. Transformations of that 'state space' act on processes. However, as we have seen, the set of all possible processes stands in one-to-one correspondence with the set of states of the physical system, since every state, as initial state, initiates a unique process, and every process starts with some initial state. So, to keep things as uncomplicated as possible (relatively speaking, of course), we apply our transformations to the physical system's state space rather than to process space. As we transform the states about, we keep in mind that each one drags around with it the process that it initiates, just like mice running around in a cage, each carrying its tail.

Then since, as we saw above, initial states are the cause subsystem and final states the effect subsystem for processes in quasi-isolated physical systems, we have that the symmetry group of the cause (see Sect. 4.4) is the group of all invertible transformations of state space that preserve initial-equivalence subspaces. And the symmetry group of the effect is the group of all invertible transformations preserving final-equivalence subspaces. And since, as follows from the discussion two paragraphs above, any transformation that preserves initial equivalence must also preserve final equivalence, we have the result:

> The 'initial' symmetry group (that of the cause) is a subgroup of the 'final' symmetry group (that of the effect).

This is the *symmetry principle for processes* in quasi-isolated systems. And in this sense we can say (see Sect. 10.7 concerning degree of symmetry):

> For a quasi-isolated physical system the degree of symmetry cannot decrease as the system evolves, but either remains constant or increases.

We call that result the *general symmetry evolution principle*. The adjective 'general' is included in the name of the principle, because the principle is derived from very fundamental considerations with no additional assumptions. Thus, it is indeed general, so general, in fact, as to make it rather useless, as we will see and remedy in Sect. 11.3. Our motive for deriving it was theoretical, not utilitarian. We should see where our fundamental considerations lead us when applied to the evolution of quasi-isolated physical systems.

11.3 The Special Symmetry Evolution Principle and Entropy

Here is the reason the general symmetry evolution principle, although perfectly valid, is not very useful. When we consider the evolution of a quasi-isolated physical system, we normally consider the sequence of states it passes through and their symmetry and are usually not interested in the entire state space of the system, in terms of whose transformations the 'initial' and 'final' symmetry groups are defined in Sect. 11.2. The symmetry group of a single state is the group of all invertible transformations of the physical system that carry the state into equivalent states, where equivalence is with respect to the laws of evolution, the laws of nature. Thus, the symmetry group of a state is just the group of permutations of all members of the equivalence subspace in state space to which the state belongs (where we generalize suitably for equivalence subspaces with discretely or continuously infinite populations, or, to use the terms presented in Sect. 8.1, of denumerably or nondenumerably infinite orders). Let us recall that by 'equivalence subspace in state space' we mean the equivalence class of states that are indistinguishable by the laws of nature, as explained in Sect. 11.2.

If the equivalence subspace contains n states, the state's symmetry group is the symmetric group of degree n, S_n, whose order is $n!$ (see Sect. 9.7). So, referring to the discussion in Sect. 10.7, the degree of symmetry of a state can be measured by the population of its equivalence subspace: The more states that are equivalent to it, the higher the order of its symmetry group, and the higher its degree of symmetry.

What can be stated about the degrees of symmetry of the sequence of states through which a quasi-isolated system evolves? Or equivalently, what do we know about the populations of the sequence of equivalence subspaces to which those states belong? In spite of the general symmetry evolution principle, we know nothing about that in general. We can, however, obtain a result by making the assumption of nonconvergent evolution: Different states always evolve into different states. Then the population of the equivalence subspace of a final state is at least equal to that of the initial state that evolved into the final state. That is because, by the equivalence principle, all members of the initial state's equivalence subspace evolve into members of the final state's equivalence subspace, and, by the nonconvergence assumption, the number of the latter equals the number of the former. Moreover, additional states, inequivalent to the initial state, may also evolve into members of the final state's equivalence subspace (convergence of equivalence subspaces), and these members will all be distinct from those we just counted. Thus:

As a quasi-isolated system evolves, the populations of the equivalence subspaces (equivalence classes) of the sequence of states through which it passes cannot decrease, but either remain constant or increase.

Or, equivalently [56]:

> *The degree of symmetry of the state of a quasi-isolated system*
> *cannot decrease during evolution, but either remains constant*
> *or increases.*

We call that result the *special symmetry evolution principle* (illustrated in Fig. 11.7). To compare the special and general symmetry evolution principles, both derive from the equivalence principle for processes, while the special principle involves the additional assumption of nonconvergent evolution.

The symmetry group of the final state, the group of permutations of all members of its equivalence subspace, clearly includes as a subgroup the symmetry group of the initial state, the group of permutations of all members of *its* equivalence subspace (all groups being considered abstractly). Actually it includes as a subgroup the symmetry group of every state evolving into a state equivalent to the final state. Further-

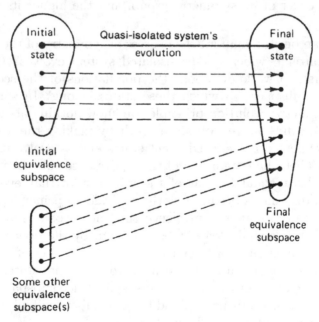

Fig. 11.7. The special symmetry evolution principle. The population, measuring the degree of symmetry, of the final equivalence subspace at least equals that of the initial equivalence subspace. That follows from the equivalence principle and the assumption of nonconvergent evolution. The solid arrow indicates the actual process being considered, while dashed arrows indicate other possible processes

more, it even includes as a subgroup the direct product (see Sect. 9.6) of the symmetry groups of all the distinct equivalence subspaces of states evolving into states equivalent to the final state (or, we could say, of all distinct equivalence subspaces converging to the final equivalence subspace).

Those statements follow from a property of permutations and symmetric groups that is easily verified, although we will not verify it here. Let the populations of the distinct equivalence subspaces that converge to the final equivalence subspace be $n_1, n_2, \ldots$. Then the population of the final equivalence subspace is $n_1 + n_2 + \cdots = n$. Recall that the group of permutations of m objects is the symmetric group S_m (see Sect. 9.7). The statements in the preceding paragraph then all follow from this relation for symmetric groups:

$$ S_n \supset S_{n_1} \times S_{n_2} \times \ldots \supset S_{n_i} \, . $$

The assumption of nonconvergent evolution (of states, not of equivalence subspaces) seems to be valid in nature for microscopic processes, at least as far as we now know. However, it is not valid for macroscopic processes. That is just fine, since the essence of the macroscopic character of a macrostate is that a macrostate is an equivalence subspace of microstates in microstate space, and such equivalence subspaces are just what we are dealing with here. They are allowed to converge and in the real world very often do.

Evolution with constant degree of symmetry is typical of microscopically considered systems, systems about which sufficient information is available so that they need not be treated by statistical methods. The mechanism of evolution with constant degree of symmetry is, as we have seen, nonconvergence of equivalence subspaces. Reversibility, or time reversal symmetry, of the laws of evolution implies evolution with constant degree of symmetry. That is easily seen: If the degree of symmetry increased during a process, the time reversed process would take place with decreasing degree of symmetry, which is forbidden by the special symmetry evolution principle. Thus, evolution with constant degree of symmetry is a necessary (but not sufficient) condition for reversibility.

As an example of evolution with constant degree of symmetry, consider a system consisting of a few nucleons and pions about which sufficient information is available. The symmetry transformations of the laws of evolution, which define state equivalence, are for this example spatial and temporal displacements, rotations, velocity boosts, particle-antiparticle conjugation, spatial (point or plane) reflection, and others. The evolution is reversible, since its laws of evolution are time reversal symmetric. Thus, the system evolves with constant degree of symmetry. As it evolves, the populations of the equivalence subspaces of the sequence of states constituting the process are all equal.

Evolution with increasing degree of symmetry is typical of macroscopically considered systems, systems about which insufficient information is available, so they must be treated by statistical methods [57]. From the special symmetry evolution principle it follows immediately that evolution with increasing degree of symmetry implies irreversibility, or time reversal asymmetry, of the laws of evolution. The mechanism of evolution with increasing degree of symmetry is convergence of equivalence subspaces, i.e., convergent evolution of macrostates (see Fig. 11.8).

An example of evolution with increasing degree of symmetry is a confined, quasi-isolated gas considered macroscopically. Let the gas initially occupy some part of the container and have volume V_i and temperature T_i. The final macrostate is a homogeneous distribution of gas in the whole container with volume V_f and temperature T_f. All initial macrostates with the same V_i and T_i, but with different parts of the container occupied, lead to the same final macrostate. And there are also initial macrostates with different initial volumes and temperatures that lead to the same final macrostate as well. The macroscopic laws of

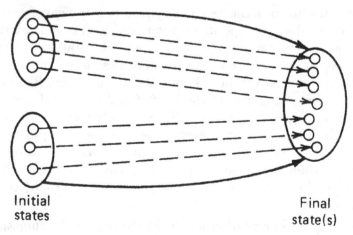

Initial   Final
states   state(s)

Fig. 11.8. Convergent evolution of macrostates, which are equivalence subspaces of microstates. Circles denote microstates, ellipses denote macrostates. Dashed arrows represent nonconvergent evolution of microstates, solid arrows show convergent evolution of macrostates

evolution are blind to the differences among microstates corresponding to the same macrostate. Thus, each macrostate is an equivalence subspace of microstate space. In microscopic terms there is convergence of equivalence subspaces during evolution. Not only do equivalent initial microstates – corresponding to the same V_i and T_i and the same occupied part of the container – lead to equivalent final microstates, but there are also inequivalent initial microstates – corresponding to the same V_i and T_i but different parts of the container, or to different V_i and T_i – that lead to equivalent final microstates. Since microstates do not converge during evolution, the population of the final equivalence subspace, which is the number of microstates corresponding to the final macrostate, is larger – actually very much larger – than the population of the initial equivalence subspace.

The final macrostate in this example is a state of stable equilibrium of the system. As such, it is the outcome of evolution from a very large number of initial macrostates. By the special symmetry evolution principle, its symmetry group, which is the group of permutations of all microstates corresponding to it, must include as a subgroup the direct product of the symmetry groups of all those initial macrostates, as we observed above, and must therefore be of relatively high order. In other words:

The degree of symmetry of a macrostate of stable equilibrium must be relatively high.

That is a general theorem, independent of the example in the context of which it was derived. Although the symmetry is with respect to permutations of equivalent microstates, it can have macroscopic manifestations. In the example the final state is homogeneous, i.e., symmetric under all permutations of subvolumes of the gas. The theorem is in accord with observation [58].

But what about those systems that insist on evolving toward reduced symmetry in spite of our theoretical arguments? For example, the solar system is thought to have evolved from a state of axial symmetry (symmetry under all rotations about its axis) and reflection symmetry (through the plane through its center and perpendicular to its axis) to the present, less symmetric state of only approximate reflection symmetry. And the evolutionary development of plants and animals seems to be toward a reduction of symmetry. The answer is that those systems do not fulfill the conditions of our argument. They are either only approximately symmetric – while our argument is based on exact symmetry – or not isolated – thus possibly subject to external influences, which we exclude – or both and are not stable against such perturbations. Since no symmetry of a macroscopic physical system is exact and no physical system is absolutely isolated, both conditions being convenient and often useful idealizations, the crucial point is that of stability. A sufficiently stable system will behave as we described, will evolve with nondecreasing symmetry, and, if macroscopic, will possess relatively highly symmetric states of stable equilibrium (see Sect. 6.2).

Under the correspondence

$$\text{degree of symmetry} \quad \longleftrightarrow \quad \text{entropy} ,$$

the special symmetry evolution principle and the second law of thermodynamics are isomorphic. Both are concerned with the evolution of quasi-isolated systems and both state that a quantity, a function of macrostate, cannot decrease during evolution. So, following good science practice, we assume that a fundamental relation exists between entropy and degree of symmetry and look for a functional relation giving the value of one as a strictly monotonically increasing function of the other. Such a relation is already known, however, from the statistical definition of entropy,

$$S = k \log W ,$$

where k denotes the Boltzmann constant and W is the number of microstates (the population of the equivalence subspace of microstate space) corresponding to the macrostate for which the value of the entropy S is thus defined. The definition holds for the case of nonconvergence of microstates during evolution. As mentioned above, W can be used to measure the degree of symmetry of the macrostate, and the function $k \log W$ is indeed a strictly monotonically increasing function of W [59, 60].

Thus entropy and degree of symmetry for quasi-isolated systems either remain constant together or increase together, with a concomitant constancy or decrease in the degree of order for the system. Note that *disorder* stands in positive correlation with symmetry. Total disorder is maximal symmetry. Organization implies and is implied by reduction of degree of symmetry. The hypothetical 'heat death' or 'entropy death' of the Universe can also be called its 'symmetry death'.

11.4 Summary

Taking the active view of transformations in Sect. 11.1, we formalized the notion of symmetry of the laws of nature, or symmetry of evolution. A symmetry transformation of the laws of nature is a transformation that commutes with the evolution transformation. We saw that the fundamental point underlying the formalism is that symmetry of evolution is an indifference of nature, whereby the laws of nature ignore some aspect of states of physical systems. We also considered the meaning of a time reversed process and time reversal symmetry of the laws of nature.

In Sect. 11.2 we saw that symmetry of the laws of nature determines an equivalence relation in a state space of a system, the equivalence of states that are indistinguishable by the laws of nature. Applying the equivalence principle, developed in Sect. 4.3, we obtained the equivalence principle for processes in quasi-isolated systems: Equivalent states, as initial states, *must* evolve into equivalent states, as final states, while inequivalent states *may* evolve into equivalent states. That led to the symmetry principle for processes in quasi-isolated systems: The 'initial' symmetry group (that of the cause) is a subgroup of the 'final' symmetry group (that of the effect). And that in turn led to the general symmetry evolution principle: For a quasi-isolated physical system the degree of symmetry cannot decrease as the system evolves, but either remains constant or increases. The latter principle is too general to be of much use.

So rather than consider the symmetry group of all of state space, in Sect. 11.3 we looked at the symmetry group of only a single state, the group of permutations of the equivalence subspace (equivalence class in state space) to which it belongs. With the assumption of nonconvergent evolution, that led to the special symmetry evolution principle: The degree of symmetry of the state of a quasi-isolated system cannot decrease during evolution, but either remains constant or increases. Or equivalently: As a quasi-isolated system evolves, the populations of the equivalence subspaces of the sequence of states through which it passes cannot decrease, but either remain constant or increase.

For systems, such as a gas, that possess nonconvergent evolution of microstates and convergent evolution of macrostates to macrostates of stable equilibrium, the special symmetry evolution principle gave the theorem that the degree of symmetry of a macrostate of stable equilibrium must be relatively high.

The special symmetry evolution principle and the second law of thermodynamics are very similar. That suggested a functional relation between entropy and degree of symmetry, which in fact is already known.

Summary of Principles

Here is a summary of the principles of symmetry that are derived in this book.

12.1 Symmetry and Asymmetry

Symmetry is immunity to a possible change.

So symmetry involves these two essential components:

1. *Possibility of a change.* It must be possible to perform a change, although the change does not actually have to be performed.
2. *Immunity.* Some aspect of the situation would remain unchanged, if the change were performed.

If these components are fulfilled, we have *symmetry* of the situation under the change with respect to the aspect. If a change is possible but some aspect of the situation is not immune to it, we have *asymmetry* of the situation under the change with respect to that aspect. If there is no possibility of a change, then the very concepts of symmetry and asymmetry are inapplicable (see Sect. 1.1).

12.2 Symmetry Implies Asymmetry

Succinctly:

Symmetry implies asymmetry.

In more detail:

> *For there to be symmetry, there must concomitantly exist asymmetry under the same change that is involved in the symmetry.*

In much more detail:

> *Symmetry requires a reference frame, which is necessarily asymmetric. The absence of a reference frame implies identity, hence no possibility of change, and hence the inapplicability of the concept of symmetry.*

Graphically:

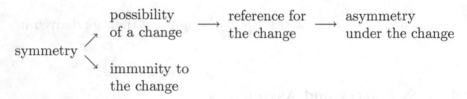

Conclusion (see Sect. 1.2):

> *For every symmetry there is an asymmetry tucked away somewhere in the Universe.*

12.3 No Exact Symmetry of the Universe

> *Exact symmetry of the Universe as a whole is an empty concept.*

Corollary:

> *For the Universe as a whole, undifferentiability of degrees of freedom means their physical identity.*

Paraphrase:

> *If it makes no difference to the Universe, then there is nothing else for it to make a difference to.*

This principle follows immediately from the conclusion of the principle 'Symmetry implies asymmetry', that for every symmetry there is an asymmetry tucked away somewhere in the Universe, as applied to the Universe as a whole (see Sect. 7.2).

12.4 Cosmological Implications

From Sect. 7.3:

1. Cosmological schemes cannot involve perfect symmetry for the Universe as a whole.
2. Cosmological schemes cannot involve fundamentally undifferentiable, yet still somehow different, degrees of freedom of the Universe.
3. Cosmological schemes with (discontinuous) phase transitions between eras cannot involve symmetry breaking.
4. High-energy physics cannot be expected to reflect precisely the situation that prevailed during earlier cosmic eras that evolved into the present era via (discontinuous) phase transitions, although it might be indicative. Specifically, any symmetry emerging at high energies cannot have been a feature of such earlier eras.

12.5 The Equivalence Principle

Roughly:

   Equivalent causes – equivalent effects.

Precisely:

   Equivalent states of a cause $\longrightarrow$ *equivalent states of its effect.*

The principle is derived from the existence of causal relations in nature and from the character of scientific laws as expressions of those causal relations. The equivalence principle is fundamental to the application of symmetry in science (see Sect. 4.3).

12.6 The Symmetry Principle

Roughly:

   The effect is at least as symmetric as the cause.

Precisely:

   The symmetry group of the cause is a subgroup of the symmetry group of the effect.

The symmetry principle, also known as Curie's principle, is derived directly from the equivalence principle and, of the two principles, is the one that is most commonly used (see Sect. 4.4).

12.7 The Equivalence Principle for Processes

> *Equivalent states, as initial states,* must *evolve into equivalent states, as final states, while inequivalent states* may *evolve into equivalent states.*

This principle is derived by applying the equivalence principle to natural evolution processes of quasi-isolated physical systems (see Sect. 11.2).

12.8 The Symmetry Principle for Processes

> *The 'initial' symmetry group (that of the cause) is a subgroup of the 'final' symmetry group (that of the effect).*

The symmetry principle for processes is derived from the equivalence principle for processes and is essentially the application of the symmetry principle to natural evolution processes of quasi-isolated physical systems (see Sect. 11.2).

12.9 The General Symmetry Evolution Principle

> *For a quasi-isolated physical system the degree of symmetry cannot decrease as the system evolves, but either remains constant or increases.*

This principle follows immediately from the symmetry principle for processes. The general symmetry evolution principle has theoretical significance, but is so general as to be quite useless (see Sect. 11.2).

12.10 The Special Symmetry Evolution Principle

Usually:

> *The degree of symmetry of the state of a quasi-isolated system cannot decrease during evolution, but either remains constant or increases.*

Equivalently:

> *As a quasi-isolated system evolves, the populations of the equivalence subspaces (equivalence classes) of the sequence of states through which it passes cannot decrease, but either remain constant or increase.*

This is the useful symmetry evolution principle. It is derived from the equivalence principle for processes with the additional assumption of nonconvergent evolution. This principle is closely related to the second law of thermodynamics (see Sect. 11.3).

References

Preface

1. B. Greene: *The Fabric of the Cosmos: Space, Time, and the Texture of Reality* (Knopf, New York 2004)

2. V.J. Stenger: *The Comprehensible Cosmos: Where Do the Laws of Physics Come from?* (Prometheus, Amherst NY 2006)

Chapter One

3. J. Rosen: The primacy of asymmetry over symmetry in physics. In: *Physics and Whitehead: Quantum, Process, and Experience*, ed. by T.E. Eastman, H. Keeton (SUNY Press, Albany 2003), pp. 129–135

4. J.N. Shive, R.L. Weber: *Similarities in Physics* (Wiley, Hoboken NJ 1982)

5. W.B. Jensen: Classification, symmetry and the periodic table. In: *Symmetry: Unifying Human Understanding*, Vol. 1, ed. by I. Hargittai (Pergamon, Oxford 1986) pp. 487–510

Chapter Two

6. J. Rosen: *The Capricious Cosmos: Universe Beyond Law* (Macmillan, New York 1991)

7. P.C.W. Davies: *Other Worlds: A Portrait of Nature in Rebellion: Space, Superspace, and the Quantum Universe* (Penguin, London 1997)

8. P.C.W. Davies, J.R. Brown (Eds.): *The Ghost in the Atom: A Discussion of the Mysteries of Quantum Physics* (Cambridge University Press, Cambridge 1993)

9. T.A. Debs, M.L.G. Redhead: *Objectivity, Invariance, and Convention: Symmetry in Physical Science* (Harvard University Press, Cambridge MA 2007)

10. P.C.W. Davies: *Superforce: The Search for a Grand Unified Theory of Nature* (Simon & Schuster, New York 1984)

11. J. Rosen: Extended Mach principle, Am. J. Phys. **49**, 258–264 (1981)

12. J. Rosen: When did the Universe begin?, Am. J. Phys. **55**, 498–499 (1987)

13. W.B. Jensen: Classification, symmetry and the periodic table. In: *Symmetry: Unifying Human Understanding*, Vol. 1, ed. by I. Hargittai (Pergamon, Oxford 1986) pp. 487–510

14. M.B. Hesse: *Models and Analogies in Science* (University of Notre Dame Press, Notre Dame IN 1966)

15. L.M. Lederman, C.T. Hill: *Symmetry and the Beautiful Universe* (Prometheus, Amherst NY 2004)

16. Y. Ne'eman, Y. Kirsh: *The Particle Hunters*, 2nd edn. (Cambridge University Press, Cambridge 1996)

Chapter Three

17. P.W. Stephens, A.I. Goldman: The structure of quasicrystals, Scientific American **280** (4), 24–31 (April 1991)

18. T. Kibble: Phase-transition dynamics in the lab and the universe, Physics Today **60** (9), 47–52 (September 2007)

19. P. Palffy-Muhoray: The diverse world of liquid crystals, Physics Today **60** (9), 54–60 (September 2007)

20. B. Greene: *The Fabric of the Cosmos: Space, Time, and the Texture of Reality* (Knopf, New York 2004)

21. V.J. Stenger: *The Comprehensible Cosmos: Where Do the Laws of Physics Come from?* (Prometheus, Amherst NY 2006)

Chapter Four

22. P.C.W. Davies: *Superforce: The Search for a Grand Unified Theory of Nature* (Simon & Schuster, New York 1984)

23. J. Rosen: Extended Mach principle, Am. J. Phys. **49**, 258–264 (1981)

24. P. Curie: Sur la symétrie dans les phénomènes physiques, symétrie d'un champ électrique et d'un champ magnétique, Journal de Physique (3rd ser.) **3**, 393–415 (1894). Reprinted in: *Œuvres de Pierre Curie* (Gauthier-

Villars, Paris 1908) pp. 118–141. English translation: J. Rosen, P. Copié: On symmetry in physical phenomena, symmetry of an electric field and of a magnetic field. In: *Symmetry in Physics: Selected Reprints*, ed. by J. Rosen (American Association of Physics Teachers, College Park MD 1982) pp. 17–25

25. R. Shaw: Symmetry, uniqueness, and the Coulomb law of force. Am. J. Phys. **33**, 300 (1965). Reprinted in: *Symmetry in Physics: Selected Reprints*, ed. by J. Rosen (American Association of Physics Teachers, College Park MD 1982) pp. 27–32

Chapter Five

26. R. Shaw: Symmetry, uniqueness, and the Coulomb law of force. Am. J. Phys. **33**, 300 (1965). Reprinted in: *Symmetry in Physics: Selected Reprints*, ed. by J. Rosen (American Association of Physics Teachers, College Park MD 1982) pp. 27–32

27. H. Stephani, M. MacCullum: *Differential Equations: Their Solution Using Symmetries* (Cambridge University Press, New York 1990)

28. H. Weyl: *Symmetry* (Princeton University Press, Princeton NJ 1952)

29. J. Rosen: *The Capricious Cosmos: Universe Beyond Law* (Macmillan, New York 1991)

30. J. Gribbin: *The Search for Superstrings, Symmetry, and the Theory of Everything* (Back Bay, Newport Beach CA 2000)

31. L.M. Lederman, C.T. Hill: *Symmetry and the Beautiful Universe* (Prometheus, Amherst NY 2004)

32. Y. Ne'eman, Y. Kirsh: *The Particle Hunters*, 2nd edn. (Cambridge University Press, Cambridge 1996)

33. P. Halpern: *The Great Beyond: Higher Dimensions, Parallel Universes and the Extraordinary Search for a Theory of Everything* (Wiley, Hoboken NJ 2004)

34. B. Greene: *The Elegant Universe: Superstrings, Hidden Dimensions, and the Quest for the Ultimate Theory* (Vintage, New York 2005)

35. L. Smolin: *The Trouble with Physics: The Rise of String Theory, the Fall of a Science, and What Comes Next* (Houghton Mifflin, Boston 2006)

36. P. Woit: *Not Even Wrong: The Failure of String Theory and the Search for Unity in Physical Law* (Basic Books, New York 2006)

Chapter Six

37. G. Birkhoff: *Hydrodynamics: A Study in Logic, Fact, and Similitude* (Greenwood, Westport CT 1978)

Chapter Seven

38. L. Smolin: *The Life of the Cosmos* (Oxford University Press, New York 1997)

39. J. Rosen: Extended Mach principle, Am. J. Phys. **49**, 258–264 (1981)

40. B. Greene: *The Fabric of the Cosmos: Space, Time, and the Texture of Reality* (Knopf, New York 2004)

41. V.J. Stenger: *The Comprehensible Cosmos: Where Do the Laws of Physics Come from?* (Prometheus, Amherst NY 2006)

42. T. Kibble: Phase-transition dynamics in the lab and the universe, Physics Today **60** (9), 47–52 (September 2007)

43. J.V. Narlikar: *The Primeval Universe* (Oxford University Press, Oxford 1988)

44. C.W. Misner, K.S. Thorne, J.A. Wheeler: *Gravitation* (Freeman, San Francisco 1973)

45. J. Rosen: *The Capricious Cosmos: Universe Beyond Law* (Macmillan, New York 1991)

46. J. Rosen: When did the universe begin?, Am. J. Phys. **55**, 498–499 (1987)

47. E. Harrison: *Masks of the Universe: Changing Ideas on the Nature of the Cosmos*, 2nd edn. (Cambridge University Press, Cambridge 2003)

Chapter Eight

48. J. Neuberger: Rotation matrix for an arbitrary axis, Am. J. Phys. **45**, 492–493 (1977)

49. A. Palazzolo: Formalism for the rotation matrix of rotations about an arbitrary axis, Am. J. Phys. **44**, 63–67 (1976)

Chapter Ten

50. J. Neuberger: Rotation matrix for an arbitrary axis, Am. J. Phys. **45**, 492–493 (1977)

51. A. Palazzolo: Formalism for the rotation matrix of rotations about an arbitrary axis, Am. J. Phys. **44**, 63–67 (1976)

52. H. Zabrodsky, S. Peleg, D. Avnir: Symmetry as a continuous feature, IEEE Transactions on Pattern Analysis and Machine Intelligence **17**, 1154–1166 (1995)

53. V. Bargmann: Note on Wigner's theorem on symmetry operations, J. Math. Phys. **5**, 862–868 (1964). Reprinted in: *Symmetry in Physics: Selected Reprints*, ed. by J. Rosen (American Association of Physics Teachers, College Park MD 1982) pp. 147–153

Chapter Eleven

54. R.P. Feynman: *The Character of Physical Law* (Modern Library, New York 1994)

55. J. Rosen: *The Capricious Cosmos: Universe Beyond Law* (Macmillan, New York 1991)

56. P. Renaud: Sur une généralisation du principe de symétrie de Curie, C. R. Acad. Sci. Paris **200**, 531–534 (1935). English translation: J. Rosen, P. Copié: On a generalization of Curie's symmetry principle. In: *Symmetry in Physics: Selected Reprints*, ed. by J. Rosen (American Association of Physics Teachers, College Park MD 1982) p. 26

57. H. Callen: Thermodynamics as a science of symmetry, Found. Phys. **4**, 423–443 (1974)

58. L.L. Whyte: Tendency towards symmetry in fundamental physical structures, Nature **163**, 762–763 (1949)

59. A. Sellerio: Entropia, probabilità, simmetria, Nuovo Cimento **6**, 236–242 (1929)

60. A. Sellerio: Le simmetrie nella fisica, Scientia **58**, 69–80 (1935)

Further Reading

Chapter One

For elementary introductions to symmetry see:

I. Hargittai, M. Hargittai: *Symmetry: A Unifying Concept* (Shelter, Bolinas CA 1994)

J. Rosen: *Symmetry Discovered: Concepts and Applications in Nature and Science* (Dover, Mineola NY 1998)

I. Stewart: *Why Beauty Is Truth: The Story of Symmetry* (Perseus, New York 2007)

More advanced introductions:

G. Darvas: *Symmetry* (Birkhäuser, Basel 2007)

H. Weyl: *Symmetry* (Princeton University Press, Princeton 1952)

Escher has given us fantastic illustrations of symmetry, which can be enjoyed in:

M.C. Escher: *M.C. Escher* (Taschen, Köln 2006)

J.L. Locker: *The Magic of M.C. Escher* (Abrams, New York 2000)

D. Schattschneider: *M.C. Escher: Visions of Symmetry* (Thames & Hudson, London 2004)

D. Schattschneider, M. Emmer: *M.C. Escher's Legacy: A Centennial Celebration* (Springer, Berlin Heidelberg New York 2003)

See also:

A. Holden: *Shapes, Space, and Symmetry* (Dover, Mineola NY 1971)

For further reading on the periodic table see:

E.R. Scerri: *The Periodic Table: Its Story and its Significance* (Oxford University Press, New York 2006)

And for general background from one of the all-time experts:

E.P. Wigner: *Symmetries and Reflections: Scientific Essays* (MIT Press, Cambridge MA 1970)

Chapter Three

On gauge symmetry and gauge theories of the fundamental interactions, see, for example:

P.C.W. Davies: *Superforce: The Search for a Grand Unified Theory of Nature* (Simon & Schuster, New York 1984)

J. Gribbin: *The Search for Superstrings, Symmetry, and the Theory of Everything* (Back Bay, Newport Beach CA 2000)

L.M. Lederman, C.T. Hill: *Symmetry and the Beautiful Universe* (Prometheus, Amherst NY 2004)

A. Zee: *Fearful Symmetry* (Princeton University Press, Princeton NJ 1999)

For symmetry and conservation see Lederman and Hill above. Some philosophical considerations are discussed in:

M. Lange: Laws and meta-laws of nature: Conservation laws and symmetries, Studies in History and Philosophy of Modern Physics **38**, 457–481 (2007)

The following is an example of an introduction to quantum theory:

W. Greiner: Quantum Mechanics: An Introduction, 4th edn (Springer, Berlin Heidelberg New York 2001)

Chapter Seven

B. Greene: *The Fabric of the Cosmos: Space, Time, and the Texture of Reality* (Knopf, New York 2004)

D. Lichtenberg: *The Universe and the Atom* (World Scientific, Singapore 2007)

Chapters Eight and Nine

M.A. Armstrong: *Groups and Symmetry* (Springer, Berlin Heidelberg New York 1991)

D.W. Farmer: *Groups and Symmetry: A Guide to Discovering Mathematics* (American Mathematical Society, Providence 1995)

F.M. Goodman: *Algebra: Abstract and Concrete: Stressing Symmetry*, 2nd edn. (Prentice Hall, Upper Saddle River NJ 2002)

D.L. Johnson: *Symmetries* (Springer, Berlin Heidelberg New York 2001)

Chapter Ten

Concerning quantum systems and symmetry in such systems, see, for example:

W. Greiner: *Quantum Mechanics: An Introduction*, 4th edn. (Springer, Berlin Heidelberg New York 2001)

W. Greiner, B. Müller: *Quantum Mechanics: Symmetries*, 2nd edn. (Springer, Berlin Heidelberg New York 1994)

Index

THE FRONTIERS COLLECTION

Series Editors:
A.C. Elitzur M.P. Silverman J. Tuszynski R. Vaas H.D. Zeh